The California Naturalist Handbook,
Second Edition

The California Naturalist Handbook

Second Edition

Greg de Nevers
Deborah Stanger Edelman
Adina Merenlender

UNIVERSITY OF CALIFORNIA PRESS

University of California Press
Oakland, California

Library of Congress Cataloging-in-Publication Data

Names: De Nevers, Greg author | Edelman, Deborah
    Stanger author | Merenlender, Adina Maya, 1963–
    author
Title: California naturalist handbook / Greg de Nevers,
    Deborah Stanger Edelman, Adina Merenlender.
Description: Second edition. | Oakland, California :
    University of California Press, [2025] | Includes index.
Identifiers: LCCN 2025011308 (print) | LCCN 2025011309
    (ebook) | ISBN 9780520412712 paperback | ISBN
    9780520412729 ebook
Subjects: LCSH: Natural history—California | Natural
    areas—California | Biodiversity—California
Classification: LCC QH105.C2 D43 2026 (print) |
    LCC QH105.C2 (ebook) | DDC 508.794—dc23
    /eng/20250605
LC record available at https://lccn.loc.gov/2025011308
LC ebook record available at https://lccn.loc
.gov/2025011309

Manufactured in the United States of America

GPSR Authorized Representative: Easy Access System
Europe, Mustamäe tee 50, 10621 Tallinn, Estonia, gpsr.
requests@easproject.com

34   33   32   31   30   29   28   27   26   25
10   9   8   7   6   5   4   3   2   1

publication supported by a grant from
The Community Foundation for Greater New Haven
as part of the *Urban Haven Project*

For all those who have looked at
a bird, a bug, or a seed and wondered—
may the tradition continue.

Contents

Acknowledgments

Thank you, California Naturalists, for providing your feedback on *The California Naturalist Handbook* to inform this 2nd edition. We also extend our gratitude to the following people for their contributions: Doug McCreary for contributing to the original forest chapter, Matt Deitch and Ted Grantham for contributing to the original water chapter, Rebecca Perlroth for additional text on geology, Carol Blaney for adding to the interpretation chapter, and David Grantz for contributing the air quality section. We also received valuable reviews from Don Hankins on our discussions of California Indigenous Peoples and from Rebecca Sarah Hernandez and Michelle Peeters on diversity and inclusion. Greg Ira provided helpful input and unwavering support for this revised book and dedicated leadership to UC Environmental Stewards. Many experienced California Naturalist course instructors provided helpful suggestions for this second edition. We are incredibly grateful to Karyn Utsumi for her exceptional assistance with so many details related to document and illustration preparation. Sandra Osterman provided expertise and graphic updates to several of the illustrations. Pamela Byrnes, Sarah-Mae Nelson, Michelle Peeters, Jill Santos, and Alexandra Stefancich provided constructive feedback and the alt text, and we are grateful to them for all they do for UC Environmental Stewards.

Thanks also to Rob Blair and Amy Rager of the Minnesota Master Naturalist program, who were so generous in sharing their time, materials, and lessons learned to help us launch the California Naturalist courses. In addition, we would like to thank the following people for their help in reviewing and contributing to this book: Peter Alagona, Heidi Ballard, Cameron Barrows, Jamie Bartolome, Chris Cameron, Steve Cardimona, Judd Curran, Katie Hardy, Luhui Isha, Stephanie Kaza, Steve Lautze, Brianna McGonagle, Peter Moyle, Seth Myrick, Prahlada Papper, Jessica Reichard, Mari Rodin, Elizabeth Salomone, Joe Scriven, Bill Tieje, and Mati Waiya.

Thanks to Louise Doucette, whose edits vastly improved the readability of the text, to Thérèse Shere for indexing, and to Naja Pulliam Collins, Sylvie Bower, Francisco Reinking, and the entire UC Press team for working so hard with us on this project. Special thanks to Chuck Crumly for his early support of this book.

The initial development of this handbook was funded in part by the California Wildlife Conservation Board and the University of California's Renewable Resources Extension Act. We extend heartfelt thanks to Julie Fetherston, Steve Barnhart, Michael (Shawn) Brumbaugh, and Rebecca Perlroth for their collaboration in piloting the California Naturalist course and providing essential feedback on the original *Handbook*.

We are indebted to our families for their support. Greg de Nevers thanks Maggie and David Cavagnaro for teaching him the power and joy of a dissecting microscope, for answering his questions with more questions, and for showing him a vision of how wild and free life can be. Susan, Orion, and Sequoya, you are the joy of my life. Deborah Stanger Edelman would like to thank Reid, Eli, and Noah for their unwavering love and encouragement. Adina Merenlender thanks her beloved husband, Kerry Heise, a knowledgeable botanist and so much more, along with their son, Noah, and daughter, Ariella, for their steadfast support and love.

Preface

California is a naturalist's paradise filled with beauty, from the granite walls of Yosemite, to the wildflowers of the Anza-Borrego Desert, from the rolling oak woodlands of the inland valleys, to the tide pools of the Pacific coast. This diversity of landscapes combined with a mild Mediterranean climate is part of what makes California so appealing to visitors and residents alike.

California is a global biodiversity hotspot and has a distinct cultural history. Despite rapid urbanization and intensive agricultural development, the state's unique and diverse ecosystems still thrive, from the stunning coastlines to desert beauty. Yes, we face the challenge of protecting this rich biodiversity amidst ongoing growth, but every Californian can contribute to positive change. Together, we have the power to ensure that California's incredible biodiversity flourishes for generations to come.

When we embarked on writing the first edition of *The California Naturalist Handbook*, the California Naturalist courses were just getting started and we had no idea how they would grow or how much we would grow with them. This second edition incorporates not only changes that have happened within the state of California, but also changes in our understanding of how those changes impact and have been impacted by the people who live here. Extreme droughts, wildfires, and floods predicted in the first edition have become all-too-common realities. In this edition, we expand on these topics and seek to incorporate a resilience-oriented lens on addressing the environmental challenges we face. Likewise, we uplift the importance of Indigenous knowledge and stewardship and discuss environmental justice.

A few comments about our use of language. Describing the original inhabitants of the region now known as California and their descendants proves challenging, as no single term encapsulates their diverse identities and cultures. We strive to honor their ongoing communities and leadership by using specific Tribe and Tribal community names whenever feasible, alongside more general terms such as *Indigenous Peoples*, *Native American*, *Tribes*, and *Tribal* when writing about communities or entities, depending on context.

In addition, throughout this text, the authors use the scientifically accepted term *Mediterranean climate* to describe the highly unusual climate that California shares with only four other regions of the world. To many, this term seems odd, since the climate is not exclusive to the Mediterranean region. We invite readers to help begin a conversation about alternatives to this term.

To address our current challenges, Californians not only must have an understanding of our natural communities and how they function but also must learn to communicate about nature to diverse interest groups, as well as experience collecting data and nurturing wild places. This understanding will inform our decision-making processes, from making wise choices about how much water to use in our households to implementing smart growth initiatives.

The California Naturalist Handbook provides science-based information about California's natural history and lays the basis for skills that naturalists use to address today's environmental issues. Although we include some information about coastal communities and nearshore currents, the book is primarily focused on terrestrial ecosystems.

It also serves as the primary text for the California Naturalist certification courses developed by UC Environmental Stewards of the University of California Agriculture and Natural Resources Division. The mission of UC Environmental Stewards is to foster a diverse community of stewards taking action to improve sustainability and resilience for California's communities and ecosystems through education and service. The California Naturalist training program promotes environmental literacy and stewardship through discovery and action. The program provides hands-on instruction on nature observation along with exposure to real-world environmental projects designed to enhance the participants' personal connection with the natural world, as well as inspire them to take collective action on behalf of the environment.

This engaged community of stewards is essential to protect biodiversity, participate in ecological restoration, and provide climate resilience throughout California. In addition, we hope that by strengthening this constituency for nature we will also raise awareness of the importance of California's institutions for higher education and its museums, field stations, and teaching programs—which advance our understanding of botany, zoology, geology, and other natural sciences to the benefit of all. Being a naturalist is important for the entire community of life, including the next generation of Californians.

We believe that part of the strength of this book comes from the different perspectives that each of us brings. Greg has 35 years of experience working as a naturalist in California and as a scientist in the New World tropics. He has a talent for pointing out significant details and a flair for storytelling. Greg's broad understanding of plant and animal communities provides the basis for the perspective this book offers California naturalists. Deborah has worked for decades managing restoration and conservation projects, as well as with community groups, government agencies, and businesses developing a wide range of environmental programs. She brought a passion for environmental policy and advocacy to the book, and her attention to detail and organizational skills kept us on track. Adina is a Professor of Cooperative Extension in Conservation Science at University of California, Berkeley, in the Environmental Science, Policy, and Management Department and is an internationally recognized conservation biologist known for land use planning, watershed science, ecological connectivity, and naturalist and climate stewardship training. Her ability to see the big picture and draw on work by experts enhanced the book throughout. We contributed different levels of effort to each chapter, depending on our expertise. However, through the process of consolidating sections and multiple reviews, we each had substantial input to all chapters and tried to merge our voices to make the book enjoyable for the reader. The power of our collaboration and dedication to California Naturalists is what made this book possible.

With this *Handbook* we invite you to deepen your understanding of the natural world around you. To enhance your learning, take a hike in your regional park, participate in a local creek cleanup, or join a community science project. Observe seasonal changes and how they affect your local open space, woodland, or stream. As your understanding of natural communities grows, share your observations with friends and neighbors. Invite them to discover the animals and plants in the area and participate in stewardship and conservation. By observing and communicating with others about California's precious natural heritage, you join a long line of California naturalists and become part of the solution to the complex challenges we face.

We encourage you to delve more deeply into each topic, in particular to explore the California Naturalist regional publications, which provide more detail on some of California's bioregions. These and other readings and resources, as well as information on classes, are available on the UC Environmental Stewards website, https://calnat.ucanr.edu/.

Finally, if there is one takeaway that we hope to leave you with, it is to go outside and explore the world for yourself. Your interactions with and as part of nature are the foundation of everything you know, everything you will learn and care about, and work to protect. This place we call California, this place we call home, is worth your attention and effort.

1

California Natural History and the Role of Naturalists

What does it mean to be a naturalist? There is often an expectation that naturalists know the names of plants and animals and their natural history, such as their preferred habitat, prey, and behavior. But in your endeavor to become a naturalist, it's not the facts you know that matter. After all, in the digital world, information is easily available. The real goal is to spend time outside observing nature wherever you are and to wonder or ask questions about what you hear, see, smell, and touch and to feel the connections between you, what you observe, and all you know about the world around you. As a naturalist you will enjoy subtle changes in the kinds of birds that are singing and what plants are flowering. Perhaps observing nature will also offer you time to think deeply about who you are, where you are, and what your connections to the Earth and its many inhabitants are.

In this book you will find lots of information, or perhaps what seems like too little information on too many things. The most important point to carry with you as you work your way through this book is that your own experience observing nature is the foundation of being a naturalist. Watching a bird you see out your window can show you a lot about the power of observation if you look at it closely enough and see not just the bird, but all it is connected to—what it eats, where it rests, who might eat its eggs? Contemplating connections takes time, but it's well worth it. Ask yourself questions: How did that insect get to this place? What does it eat? Have I seen something similar before, and if so, where?

California is an incredible place to be a naturalist. For people who like to spend time outside, exploring unique places and sharing their favorite trail or rare species with others, the opportunities in California abound. The variety of landscapes provides diverse and unique living laboratories for aspiring naturalists. We have our nation's oldest lake, lowest point, and tallest trees and seemingly endless opportunities for learning, action, and stewardship. From spider watching in your neighborhood to exploring the far corners of the state, the opportunities provided by California's ecosystems are truly extraordinary.

Artifacts dating back over 18,000 years have been found in some areas of California, and creation stories recognize that Indigenous Peoples stewarded the land since time immemorial. What we now call California supported some of the most densely populated regions of earlier North America, with human populations estimated to be in the range of 300,000 to 1 million living in local communities and family bands. Given what the food resources and conditions were before the arrival of foreign diseases, there may have been closer to a million people living in hundreds of small communities with unique cultures and languages influenced by the local environment and food availability. Land, water, and foods characteristic of diverse natural ecosystems are vital to over 200 Tribes and Tribal communities in California today. Land stewardship is foundational to these communities' traditional practices, and nature provides sustenance, spiritual connection, and overall well-being.

CALIFORNIA'S BIODIVERSITY

California is one of the most diverse places on Earth, with an astonishing number of native species: 27,000 terrestrial invertebrates, 70 freshwater fish, 68 amphibians, 100 reptiles, 429 birds, 185 mammals, and more than 6,500 plants. Many of these species are found only in California. California has huge variations in topography and climate, with dramatic mountain ranges, valleys, and deserts where distinct natural communities have evolved. These variations result in 10 different bioregions, each with its characteristic drainage, topography, climate, and habitat types. California harbors such a wide variety of habitats and species that it is recognized as a global biodiversity hotspot.

Biodiversity is the diversity of life found at all hierarchical levels—from genes, species, and communities to entire ecosystems. In California, high levels of biodiversity are in evidence everywhere, from tiny mosses to giant redwoods. California has the largest number of endemic

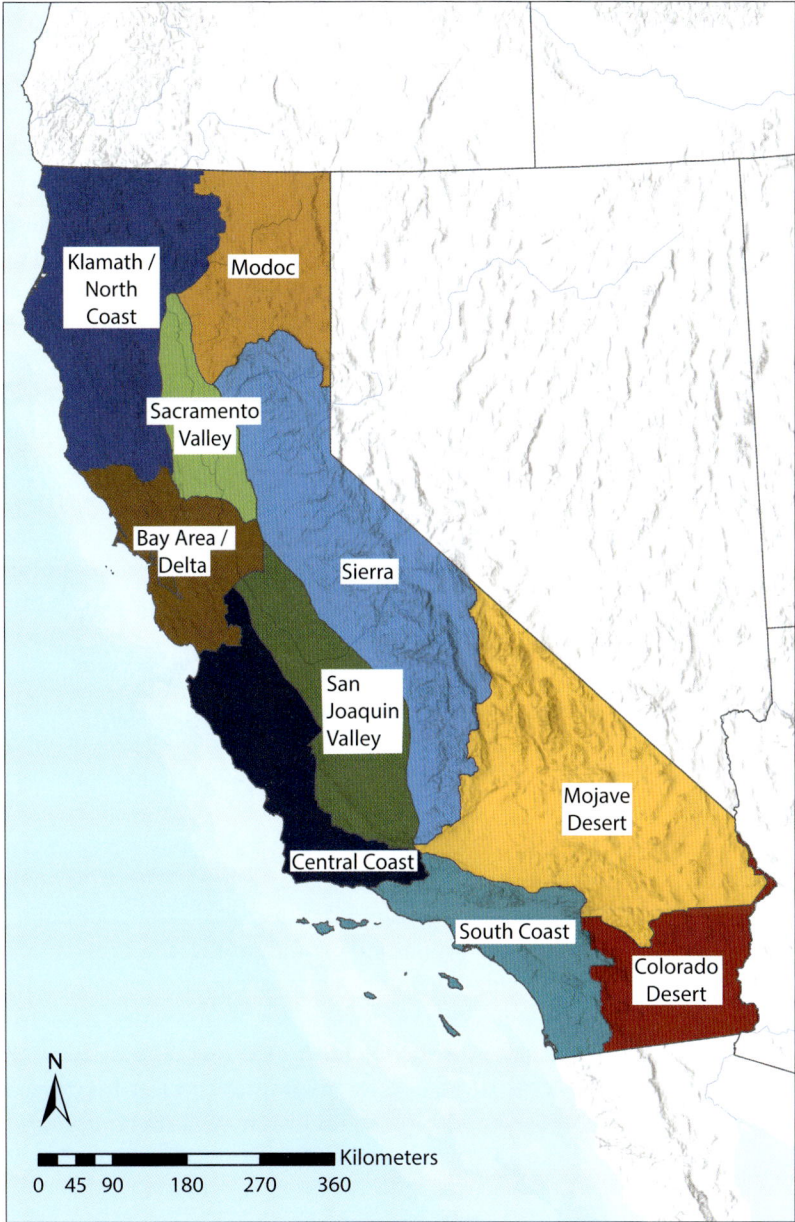

The bioregions of California are distinguished from one another by their geology, climate, topography, and associated plant and animal communities. Courtesy of UC ANR IGIS with source map from CalFire: Fire and Resource Assessment Program.

species of any state. *Endemic* is a term that indicates that a species is uniquely found within a specific geographic range. A species can be endemic to Solano County, like the delta green ground beetle, which lives in Solano County and nowhere else on Earth. Another species can be endemic to California, like the blue oak, which is widespread in California but grows in none of the other 49 states or on other continents. A species can be endemic to the New World, like the puma and the bighorn sheep. On the other hand, some taxa are cosmopolitan. For example, the bracken fern (*Pteridium aquilinum*) can be found on all continents except Antarctica. In sum, *endemic* is usually used to refer to species that have limited geographic ranges.

California contains a variety of topographical and physical features that result in wide variation in temperature, rainfall, and soil type that has led to the evolution of species found nowhere else. These factors, coupled with the state's large size, result in its 10 distinct bioregions. This makes California an interesting place to explore biogeography and learn how species are distributed within and between these bioregions and at what abundance. Most other states are far more homogeneous and hence encompass fewer bioregions. Minnesota, for example, has 3 bioregions. With so many unique species and ecosystems in California, human-induced change has broad implications for its flora and fauna. For example, when valley oak woodland is converted to a housing development, it can be difficult to find similar habitat nearby to substitute for what was lost. If the soils, slope, topography, temperature, and water availability of another place are outside the range tolerated by valley oak trees, then a valley oak woodland cannot be newly established there. With valley oak woodlands occupying only a portion of their historical range, continued loss of these woodlands puts this natural community at risk.

Most people in California live in urban areas where they can find a surprising number of species. Parks, restored creek corridors, gardens, and other green spaces can provide habitat for the more urban-adapted species and may serve as stopovers for migrating species. Improving habitat in urban areas is important for biodiversity conservation and human well-being.

A TURNING POINT FOR BIODIVERSITY

Throughout Earth's history, species have faced extinction events repeatedly, as well as an ongoing ebb and flow of their numbers; however, current anthropogenic influences, that is, those caused by people, have accelerated the extinction rate far beyond natural background levels, posing a significant threat to biodiversity. If the current rate of biodiversity loss continues, we will experience the most extreme extinction event of the past 65 million years.

Biodiversity loss is in large part a consequence of habitat loss and ecosystem degradation, which are in turn caused primarily by land use change. The Wildlife Conservation Society has calculated that the human footprint—land use change or other evidence of human influence—is detectable across 83 percent of the land area in the world, excluding Antarctica. An example of what can result from extensive land use change is Southern California, where as much as 90 percent of the historic riparian habitat has been lost to agriculture, urban development, flood control, and other alterations. Synergistic effects of habitat loss, fragmentation, and global warming can amplify biodiversity decline.

The expected loss of species due to continued land use change will impact humans because we all depend on natural ecosystems. Many of our medicines, foods, and fibers—indeed, the basis for our economies and survival—come from plants and animals. Biodiversity and natural processes are responsible for maintaining air quality, soil productivity, and nutrient cycling; moderating climate; providing fresh water, food, and pollination; breaking down pollutants and waste; and controlling parasites and diseases—the essentials of what we need to live on Earth.

The interdependency of humans and other species is important to recognize, and it provides strong justification for conservation in the interest of protecting humanity. At the same time, the intrinsic value of nature is equally important, and many are called to care for natural ecosystems and prevent extinction. Habitat loss and extinction result in lost opportunities for personal inspiration and cultural enrichment, whether by bird-watching, harvesting wild food, or simply enjoying a scenic view. To become stewards of the Earth, we need to acknowledge our interdependence with nature and our ethical, practical, and moral responsibility to prevent damage to the Earth's systems.

California, along with Hawaii, leads the nation in number of endangered species. Typically, one of five of the species you will come to know in California has been declared endangered, threatened, or "of special concern" by state or federal agencies. Some of these species are protected under the US or the California Endangered Species Act, which both list species known to be in serious danger of becoming extinct and make it illegal to take or harm these species through habitat loss or degradation.

The California condor, the largest land bird in North America, is a scavenger that has persisted since the days of the megafauna. Photo by Ram Vasudev.

Unique California: Largest, Oldest, Hottest . . .

California is unusual in so many ways that it is difficult to enumerate them all. Here are a few fun facts.

OLDEST, LARGEST, TALLEST

- Clear Lake is the largest lake entirely within California and among the oldest lakes in North America.
- Methuselah, a bristlecone pine (*Pinus longaeva*) in the White Mountains, is recognized as one of the very oldest living trees in the Western Hemisphere. It is nearly 5,000 years old.
- The General Sherman tree, a giant sequoia (*Sequoiadendron giganteum*) in Sequoia National Park, is estimated to be 275 feet tall and is over 36 feet in diameter, making it the largest (by volume) tree in the world.
- The tallest trees in the world are the coast redwoods (*Sequoia sempervirens*) along California's north coast.
- San Francisco Bay and Delta together make up the West Coast's largest estuary. In the entire country, only the Chesapeake Bay is larger.

NATURAL EXTREMES

- The lowest point in North America, at 282 feet below sea level in Death Valley, is located less than 100 miles from 14,505-foot Mt. Whitney, the highest point in the contiguous United States.
- Death Valley also has the hottest and the driest points in North America.
- California has all three kinds of tectonic plate boundaries (divergent, convergent, and transform).
- California is one of only five regions in the world with the climate type referred to as Mediterranean, characterized by the long dry summers and wet winters.

(continued)

INTRODUCTION TO BEING A NATURALIST

Naturalists observe, study, and interpret the natural world and engage in land stewardship. Humans have always been observers and natural historians by necessity. Just to survive, we have had to observe, measure, speculate about, and communicate about the world. Indigenous Peoples lived,

Unique California: Largest, Oldest, Hottest . . . *(continued)*

BIODIVERSITY

- The California Floristic Province has been designated a biodiversity hotspot by Conservation International. Like the other hotspots worldwide, it has lost at least 70 percent of its original habitat and contains at least 1,500 species of endemic vascular plants.
- More than half (63%) of California's native freshwater fish species are endemic.

CALIFORNIA COMMUNITIES

- California is the most populous state in the United States, with over 39 million people in 2020—one out of every eight Americans lives in California.
- Three of the 15 most populated US cities are in the state of California: Los Angeles, San Diego, and San Jose.
- California supported the most densely populated areas of Indigenous Peoples precontact in what we now call the United States.
- There are 109 federally recognized Native Tribes, about 45 formerly recognized Tribes that were terminated under the US termination policy, and many Tribal communities that were never recognized by the federal government; they are deeply connected to their homelands and dedicated to culture and Indigenous stewardship.
- California has the largest economy of any state in the country and in 2025 was the fourth-largest economy in the world.
- California grows nearly half of the fruits, nuts, and vegetables for the entire country. Almost all almonds, artichokes, dates, figs, kiwis, olives, persimmons, pistachios, prunes, raisins, wine grapes, and walnuts bought in the United States are grown in California.

and in some areas still live, deeply aware of their surrounding ecosystems, relying on observing, knowing, stewarding, and teaching the next generation about nature. From the most seemingly depauperate mountaintop to the most diverse shrubland, people have discovered foods, medicines, poisons, useful materials, and aesthetics in the ecosystems they inhabit. The knowledge held by Indigenous Peoples is critical for stewardship purposes and astonishing in its breadth and depth.

The Endangered Species Act

The federal Endangered Species Act (ESA) was passed in 1973 to prevent species extinctions by protecting threatened and endangered species and their habitats. The ESA defines an endangered species as one "in danger of extinction throughout all or a significant portion of its range."

Here are some better-known recipients of ESA protection:

- The American peregrine falcon increased from 324 pairs in 1975 to 1,700 pairs in 2000 and was removed from the list in 1999. There are now an estimated 3,000 breeding pairs in North America.
- The gray wolf, once abundant across North America, decreased in numbers due to widespread extermination efforts and gained federal protection under the ESA in 1974. Thanks to an active recovery program, wolves are now breeding in northeastern California and expected to spread.
- The California condor population had declined to only 22 individuals by 1982! The condor is currently under intensive management and of 2022, the wild condor population numbered 347.

The state of California has 292 federally listed species. There is also a California Endangered Species Act (CESA) that protects approximately 250 species. The California Natural Diversity Database provides an inventory of the status and locations of rare, threatened, and endangered plants and animals in California. There are some limitations to both the federal and state laws:

- Species are not listed until population numbers are dangerously low. Many biologists would like to see protections in place before numbers get perilously small.
- Political maneuvering and low levels of funding can affect how these laws are enforced.
- Since species and their habitats are protected on both public and private land, some landowners are hostile toward the Act and ignore the regulations.
- Focus on protection of a single species may no longer be possible where many species are listed, so attempts are being made to protect entire natural communities through habitat and natural community conservation plans.

Northern Valley Yokuts
Miwok Washoe
Mono Lake Northern Paiute
Ohlone (Costanoan)
Foothill Yokuts
Owens Valley Paiute-Shoshone
Western Mono (Monache)
Salinan
Western Shoshone (Panamint)
Tubatulabal
Southern Valley Yokuts
Kawaiisu
Southern Paiute
Chumash Kitanemuk
Tataviam
Serrano
Mojave
Gabrielino (Tongva)
Juaneño (Acjachemen)
Cahuilla
Luiseño
Cupeño
Chemehuevi
Halchidhoma
Quechan
Kumeyaay (Diegueño/Kamia/Ipai/Tipai)

Tolowa
Yurok Karuk Shasta Modoc
Wiyot Chilula
Hupa (Hoopa Valley/Tsnungwe)
Whilkut Chimariko Achumawi (Pit River)
Mattole Wintu (Northern Wintun) Northern Paiute (Paviotso)
Lassik Atsugewi (Hat Creek)
Sinkyone Nongatl Yana
Wailaki Nomlaki (Central Wintun)
Cahto Yuki Maidu (Mountain/Northeastern)
Konkow (Northwestern Maidu)
Pomo
Lake Miwok Nisenan (Southern Maidu)
Wappo Washoe
Patwin (Southern Wintun)
Coast Miwok
Miwok
Mono Lake Northern Paiute
Northern Valley Yokuts
Ohlone (Costanoan)
Foothill Yokuts
Owens Valley Paiute-Shoshone
Southern Valley Yokuts Western Mono (Monache)
Salinan
Esselen
Tubatulabal
Western Shoshone (Panamint)
Kawaiisu

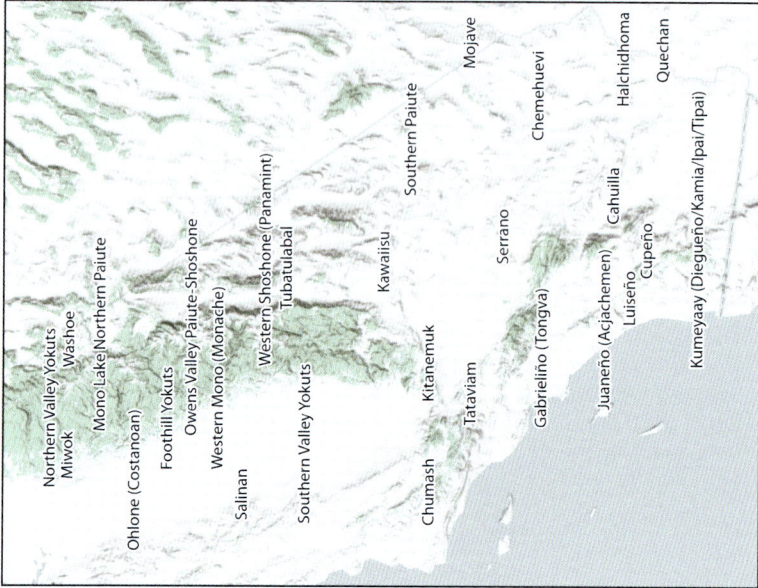

Native American cultural groupings throughout California. This map provides an overview but does not fully represent the diversity of Tribal groups in the state, nor does it necessarily reflect Tribal perspectives on ancestral or current Tribal land. Created by UC ANR IGIS using data from the Digital Atlas of California Native Americans provided by the California Native American Heritage Commission.

INDIGENOUS PERSPECTIVE AND STEWARDSHIP

We had many relatives and we all had to live together, so we'd better learn how to get along with each other. She (my mother) said it wasn't too hard to do. It was just like taking care of your younger brother or sister. You got to know them, find out what they like and what made them cry, so you'd know what to do. If you took good care of them, you didn't have to work as hard. Sounds like it is not true, but it is. When that baby gets to be a man or woman, they are going to help you out.

You know, I thought she was talking about us Indians and how we are supposed to get along. I found out later by my older sister that mother wasn't just talking about Indians, but the plants, animals, birds—everything on this Earth. They are our relatives, and we better know how to act around them or they'll get after us.

—Lucy Smith, Dry Creek Pomo

There are many ways of learning about the world, many ways of talking about the world, and many ways of thinking about the world. The Western scientific approach, in which possible explanations for observations are proposed and then tested against evidence, is one way of gaining knowledge. Indigenous Peoples' perspectives often focus on their relationships with the natural world, their spiritual beliefs, and ways of understanding the interconnectedness with the entire community of life. This Indigenous knowledge includes stewardship practices that have sustained people on the land since time immemorial and is sometimes referred to as traditional ecological knowledge.

Indigenous Peoples all over the world are expert naturalists with remarkable observations and insight into the natural processes, habitats, and organisms with which they share a long history. California Native Peoples likewise have a deep-time perspective and experience observing, talking about, and depending on California's biota.

Native Americans traditionally pass knowledge from generation to generation experientially and orally. A mother will take her daughter out to dig roots for basket making. The daughter will be in the habitat, feel the sand, sweat in the heat of the day, and struggle to pull the roots up. The mother might also talk about the activity, give verbal instruction about technique, or pass on customary knowledge associated with the experience. The words the daughter hears might be in the form of literal information or in a metaphoric form: a traditional story, a song, or a recollection. And designs for baskets and other instructions are often said to be transmitted through shared dreams.

The experiences and observations contained in this ecological knowledge represent a wealth of important information, including

Hopland Band of Pomo Indians, 100 Year Anniversary. Photo provided by the Hopland Band of Pomo Indians.

information about what plants and animals can be eaten, when they are available for harvest, and how to collect and process them—which varies depending on location, environmental conditions, and culture. California Tribal knowledge bearers know and practice Indigenous stewardship focused on interventions that improve ecological health and productivity, including the use of fire to promote habitat diversity and to encourage the production of useful animals and plants.

Indigenous cultures often emphasize the relationships between humans and the natural world. In part this is because traditional foods and cultural practices rely on natural products that often can be harvested only from intact natural communities. Practices that tie people to the land form the basis for a strong sense of place and responsibility for the world around us. Plants, animals, and physical elements such as water and mountains are seen not as mere commodities for human use, but as sacred. Since nature has spirit, using natural resources is seen as an exchange, and hence common practice is to only take what is needed and to honor what is taken. Many Tribes and Tribal communities have ceremonies, dances, and songs that are associated with world renewal religions and relate to what provides sustenance, such as acorns, strawberries, and salmon.

Native Peoples across California protect and enhance natural resources through active stewardship practices, such as distributing bulbs while harvesting, and plucking grasses rather than uprooting them—preserving them for future generations. Time-tested practices have proven to be beneficial today for native plant restoration and regeneration efforts to enhance biodiversity across many habitat types. After all, they have been nurturing plants for thousands of years, and this long history had an important role in shaping the natural communities we see in California today.

The use of fire is one area where Indigenous Peoples have a highly developed and nuanced understanding. While specific practices may vary by Tribal community, they used fire to drive and concentrate game, to open paths for travel, to alter habitat mosaics, to protect against enemies, and to safeguard villages from fire, as well as to stimulate the production of specific plants and plant parts. By burning certain chaparral stands, they stimulated new growth, and by coppicing plants such as redbud (*Cercis occidentalis*), they encouraged straight new shoots to emerge. This new growth of long, unbranched redbud stems is highly valued for design elements in coiled baskets.

It is important to keep in mind that Native Californians represent a huge variety of cultures and hundreds of languages. Once again, California tops many other places with its diversity—in this case, Indigenous cultural diversity. Even groups that identify with the same name today often include multiple languages and cultures. For example, people living in the area we now refer to as Pomo land once spoke at least seven distinct languages that are thought to share a common origin. Today they use unique names, mostly connecting to place, for their communities. For example, people living near the town we call Hopland today are Sho-Ka-Wah or "east of the river" in the Central Pomo language.

Tribes are present in all parts of California, from the coast to the mountains to the deserts. Tribal life has always been intimately tied to place. For instance, coastal Tribes like the Chumash and Miwok relied heavily on intertidal resources like shellfish. In contrast, a little farther inland, Tribes such as the Pomo and Yokuts focused more on salmon and acorns. The Timbisha Shoshone, whose ancestral lands are in what is now called Death Valley National Park, moved upslope into the mountains during the hot summers to harvest pine nuts, returning to the warmer valley during the winter. Access to different areas and ecosystems is important to Tribal life today.

Fortunately, important traditions and stewardship practices survive within Tribal communities today after years of oppression at the hands of early colonizers. The devastating genocide of California Native Peoples started in 1769 with the first European colonists and accelerated as the Catholic missions were established in the late eighteenth century. Many Indigenous people were coerced or forcibly taken from their communities to missions. The enslavement of Native Californians as a labor force in the agricultural and construction activities of the missions resulted in countless deaths due to exhaustion, malnutrition, and harsh working conditions. Crowded conditions and exposure to foreigners led to the spread of deadly diseases, such as smallpox, measles, and influenza, that disproportionately impacted Indigenous Peoples who had no immunity to these illnesses. During this period, Native Californians were often forced to convert to Christianity and prevented from speaking their own languages, which disrupted traditional ways of life, social structures, and cultural practices.

They were also persecuted by Euro-American settlers seeking timber, gold, grazing land, and agricultural land, often fueled by the belief in manifest destiny. The pursuit of land and resources was relentless and involved deliberate displacement of and violence against Indigenous Peoples. California's malign and misnamed 1850 Act for the Government and Protection of Indians legalized bounty hunting of Indigenous people, institutionalized nonrecognition of California Tribes, and legalized suppression of cultural practices. The suppression of language and culture was also institutionalized throughout federal Indian boarding schools. The legacy of these policies continues to impact Tribal communities.

The arrival of Europeans had consequences for the land as well. Introducing plants and animals led to the spread of exotic grasses and forbs, transforming California's ecosystems and reducing the number and productivity of food plants. Landscapes that had been carefully stewarded by Tribes for millennia were now subjected to new and often ecologically inappropriate management regimes.

Though this book takes a Western scientific approach to looking at California natural history, we draw inspiration from the resilience of Tribal communities across California and their Indigenous knowledge and stewardship practices.

**Establishing Local Ecological Literacy:
Where Are You Now?**

- What watershed do you live in?
- What stream is closest to where you live?
- Does it have water in it all year long, or seasonally? If seasonally, when does it go dry?
- What is the name of the nearest mountain range?
- What kinds of trees are growing near to where you live? Other plants? Insects?
- Describe where you live, in terms of both natural resources and the built environment.
- What did your neighborhood look like 100 years ago? 500 years ago?
- Who are the Indigenous Peoples whose traditional homelands you are residing in?

NOTABLE NATURALISTS

Naturalists are generalists in the best sense—cross-disciplinary, with knowledge of a system as a whole, not just the pieces. Before the formal fields of ecology, entomology, geology, zoology, and others emerged, it was naturalists who gathered the foundational observations that led to these sciences. They collected specimens, recorded observations, and became experts in certain habitats or species. Their work and collections were central to the origins of natural history museums and to the development of the theoretical underpinnings of current scientific thought. At the same time, natural history museums inflicted significant harm on Indigenous peoples. During the nineteenth and twentieth centuries, naturalists aligned with colonial powers often stole and mistreated objects, including human remains, and labeled them with derogatory terms. This exploitation caused enduring cultural and psychological damage to Indigenous communities.

Today, naturalists form a key link between scientific and common public understanding. By becoming a naturalist, you are taking a place in an important tradition of knowledge keepers. Naturalists make the world accessible to all and bring out our interest in and wonder at nature.

Charles Darwin, one of the most notable naturalists, transformed science through his publication *On the Origin of Species*. Drawing on data from his worldwide observations, collections, and extensive reading, he described and provided evidence for the theory of evolution by natural selection in 1859. Alfred Russel Wallace independently formulated the theory and, through meticulous documentation of species distributions, helped establish biogeography—the study of how and why organisms are distributed across the Earth over time. John Muir, another prolific naturalist, was born in 1838 in Scotland but moved to the San Francisco area in 1868. He developed an experience-based appreciation for the interconnectedness of nature. In *My First Summer in the Sierra*, he wrote: "When we try to pick out anything by itself, we find it hitched to everything else in the universe." Muir loved to roam the mountains, drawing and making notes as he went. He authored several books about the geology and natural history of the Sierra Nevada, turning his observations into writings that stirred people to visit and protect wild areas and helped start the Sierra Club as a way for people to work together for the preservation of wilderness. These men shared the European colonial perspectives of the time and disregarded Indigenous Peoples and their knowledge of and relationships with nature.

Many strong women pioneered aspects of California natural history. Dr. Katherine Siva Saubel, a Cahuilla elder and scholar, dedicated her life to preserving and transmitting traditional ecological knowledge of the Cahuilla People. Born in 1920, she became a leading ethnobotanist, linguist, and cultural ambassador. Her published works, including *Temalpakh: Cahuilla Indian Knowledge and Usage of Plants*, made valuable contributions to the conservation of Indigenous wisdom.

Alice Eastwood was a leader in the documentation of the flora of western North America. She identified and named 395 plant species, worked tirelessly and successfully to preserve California's redwoods, and left a legacy of male students in her wake.

Annie Alexander was a bold, gun-toting explorer in the early 1900s, and her extensive collections enriched the fields of paleontology and vertebrate zoology. She was a founder of the Museum of Vertebrate Zoology and the Museum of Paleontology at UC Berkeley and was a financial benefactor of those museums for more than 40 years.

From the mid-1800s to the mid-1900s, the botanist Katharine Brandegee traveled widely by horseback and mule, making noteworthy contributions to the study of the flora of California. In 1883 she became the first woman curator of botany at the California Academy of Sciences.

Participants in the Tribal California Naturalist Course offered by the California Tribal College in collaboration with UC Environmental Stewards journaling at UC Berkeley's Sagehen Creek Field Station. Photo by Adina Merenlender.

NOTEBOOKS AND JOURNALS: ESSENTIAL RECORDS FOR NATURALISTS

While Indigenous Peoples and early explorers laid the foundation of what we know today, there is still a lot to explore and discover about the natural world. Describing what you observe is an important part of discovery, and keeping a naturalist journal and a field notebook is one of the best ways to learn. A journal is a place to note your experiences, frame your questions, and check your facts. Journals are usually written after the fact, at home after a day in the field. Some naturalists prefer to keep only a small field notebook with them while out in nature and to enlarge field observations with additional facts and gleanings from reading in a more complete naturalist journal at a later time. Journaling is a way of setting aside time to put your thoughts in order, to frame your experience, to research questions you developed, to help you share with others, and to prompt you to be exact in your observations.

Darwin, as well as Alexander von Humboldt, wrote at night aboard ship after a day of exploring and left us journals that provide important

Nature journaling retreat at Point Reyes National Seashore. Photo by Elizabeth Kelley Gillogly, wildwonder.org.

recorded observations that we value highly today for the information they contain about a time we cannot return to. Henry David Thoreau's journal entries on blooming times of flowers around Walden Pond are proving to be useful to climate scientists today. A simple thing like noting spring's first lilac bloom in your yard allows you to check, when your friend mentions that the lilacs are blooming early (or late) the following year.

One reason for keeping a naturalist journal is that it emphasizes and explicitly values your own experience. It is a place to say "today I saw the first western kingbird of my life!" Every time you venture out into the natural world, you will have many experiences, at least some of which will be interesting or new to you. Write these down. It is critical that you value your own experience. It doesn't matter how many people before you have seen a western kingbird. In fact, it is not only the first time you see something that it is special. Every time you see a hummingbird visit a paintbrush flower is special, if you imbue it with value.

Many people like to draw in their journals. Try taking a few colored pencils along to draw some particularly appealing moss or the pattern of facial stripes on a bird. Drawing can be a wonderful way to encourage yourself to slow down, to take the time to observe nature more closely, and help you remember and savor the experience.

We recommend starting with a variation of the Grinnell method that emphasizes the collection and organization of field observations for scientific investigation. This method was developed by Joseph Grinnell (1877–1939), a noted field biologist and the first director of UC Berkeley's Museum of Vertebrate Zoology. For over a century, field biology and natural history students at UC Berkeley have been trained in this method, and it is probably the method most used by professional naturalists. While we recommend this general format because it provides a good framework for making and recording observations, this journal is yours, and over time you may modify or develop new techniques that work best for you. The important point is to use your journal to sharpen your skills of observation by recording details, and make it your own learning experience. Here are some guidelines on how to keep a field notebook and a journal based on the Grinnell method.

Field Notebook

A field notebook is the one you carry with you when you are out of doors. Use it for recording very short notes about your observations. This information will often appear in your field notebook:

· Time and date
· Location (with arrival and departure times)
· Route traveled
· Weather (including temperature, wind, precipitation type, cloud cover)
· Habitat/vegetation type (woodland, grassland, wetland)
· Species, rocks, or other natural objects seen
· General observations and comments, including human-caused disturbances and changes, both throughout the day and since a previous visit
· Drawings, maps, photos (with digital photo numbers)

Naturalist Journal

Some naturalists also keep a journal where they enlarge the short notes from their field notebook. Each field day should have a separate journal entry that includes a written description of the day's observations. A species account (if any) and a catalog of collected specimens (if any) can also be included with the day's entry. If you write your naturalist journal on a computer, it becomes searchable, which is great when the

Tools of the Trade

The great thing about being a naturalist is that it only requires a piece of paper, a pencil, and your time. However, certain items are helpful when you are observing nature: map, binoculars, camera, magnifying lens, pocketknife, waterproof notebook and pencil (pens won't write on wet paper, and ink runs), small first aid kit, sunscreen, hat, water, snack food, layers of clothing, and a watch.

Additional items may also be helpful, especially if you are leading a group: field guides, cell phone, star charts, flashlight, laser pointer, pocket mirror, clear vials for showing specimens to others, tape measure, flagging tape, an old toothbrush for cleaning fossils or rocks, small shovel, spoon, colored pencils and paper for artwork, duct tape, name tags, butcher paper and charcoal for rubbings of trees or signs, a magnet, and matches or a lighter.

journal becomes 300 pages long and covers 20 years and you want to remember when you first saw *Pseudotrillium*!

General Journaling Practices

Here are some specific points that can help format a journal:

- A bound journal with acid-free (long-lasting) paper is preferable.
- Use a pencil or a pen with water-resistant ink.
- Put your name and the year in the upper left corner of each page.
- Date each entry.
- Avoid abbreviations, as it is easy to forget these, and other readers may not be able to interpret them.
- Give a general account of the day's events, including where you went, the conditions, what you saw, and any additional thoughts or comments.
- Underline scientific names (Genus species) with straight lines, and use wavy lines for common names; this helps in scanning your journal for observations.
- Include maps (pasted into your journal or drawn) if they help with the descriptions.

Baker's meadowfoam
Limnanthes bakeri
↙ 7 transparent veins per petal
10 stamens with yellow anthers & 5 parted stigma
↑7mm
basal lf.→
bud pendulous, upper leaves w/fewer leaflets

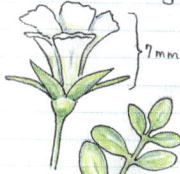

G. Hulse-Stephens
4/2/12   15:30
arr: 15:15   depart: 15:50
location: Hearst Willits Rd.
in field east of Berry Creek
crossing on north side of
road on private property.
route: take Hwy 101 to Willits,
go east on Commercial St., turn
left on Hearst Willits Rd,
continuing to where road crosses
Berry Cr.
weather: Overcast, about 49°,
              no wind
habitat: Wet meadow, growing
       in swale and road
       ruts near well with
       semaphore grass,
       Pleuropogon californicus var.
       davyi

Format for a field notebook. Courtesy of Geri Hulse-Stephens.

Species Account

This serves as a record of a specific observation of interest, usually taken repeatedly in a single place to document the ecology or behavior of a specific species; start with the name of the species, location, and date/time.

Catalog

This is a record of any specimens that you collect. Make sure to label or tag each specimen. Include the following information: your name, catalog number, date collected, collection location, and identification of the specimen, if known.

THE LANGUAGE OF NATURALISTS

The scientific community has developed a language all its own to clearly and accurately identify organisms. With about 390,000 species of vascular plants identified on Earth, we can't simply refer to the one we want to discuss as "that plant." Also, with hundreds of new species described and cataloged in museums each year, new names are needed all the time.

The problem is even more overwhelming with insects, where the numbers are truly phenomenal. The number of insects that have been described by scientists is around a million, and these are just the ones we have identified. There are probably 9 million more insect species out there, making insects the most diverse group of animals on Earth!

To allow us to communicate clearly about organisms, a naming convention was developed in the 1700s that is still very helpful today. It assigns a unique scientific name, based on Latin or Greek, to each species. This allows people from all over the world and speaking different languages to refer to the same species by the same name. For example, the scientific name for what some Pomo Peoples called *Ka-Doose* and what the Coast Miwok called *tu'-se* is *Procyon lotor*. In English we call it a raccoon. Likewise, we use *cat* in English and *gato* in Spanish, but Spanish and English speakers can be sure they are referring to the same animal when they use *Felis catus*. Some common names employed by naturalists can provide valuable descriptors, like western bluebird, or some are just fun to say, like chuckwalla.

Linnaeus's Classification System

The system scientists have devised is elegant in its simplicity and formality. It is based on the classification system created by Carl Linnaeus (1707–1778). Linnaeus was one of the first people to attempt to list all the plants and animals known at the time, to give each a unique name, and to classify them in a coherent manner. At the time of Linnaeus, Latin was the language that scientists used to communicate with one another across the globe, so Linnaeus and his contemporaries used Latin when classifying species. This tradition carries through today as scientists continue to use Latin and some Greek for genus and species names.

Each species is assigned a two-part name, a binomial. The first word of the binomial designates the genus to which the organism has been assigned, while the second designates the species. Several of these binomials have made it into popular usage. For example, *Tyrannosaurus rex* is the binomial that designates a well-known large, extinct carnivorous dinosaur. *Tyrannosaurus* is the genus name and the first half of the binomial, and the second half is *rex*—the species name.

One of the useful aspects of this naming system is that each species has been assigned a binomial that is unique. Thus, human beings are the only species called *Homo sapiens*, which translates to "knowing man." Another advantage is that the naming system reflects relationships among species. For instance, the house cat is *Felis catus*, but it is not the only member of the genus *Felis*. In fact, the genus *Felis* is global in distribution and contains about 28 species. The fact that all have been assigned to the genus *Felis* indicates that the people who have studied the relationships of the mammals think that all 28 species share a recent common ancestor and thus can be considered closely related. The same relationship is true for their prey the rats, which total more than 60 extant species, with the best known being the black rat (*Rattus rattus*) and the brown rat (*Rattus norvegicus*). Less-related ratlike species, like two newly described tweezer-beaked hopping rats from the highlands of the Philippines, have been assigned the genus name *Rhynchomys*. They are named *Rhynchomys labo* and *Rhynchomys mingan*, indicating that they are closely related to each other but less related to members of the genus *Rattus*.

The scientific system of naming, starting with the binomial that names each species, is a hierarchical system. It is designed to indicate nested levels of relationship. Just as we recognize close family relationships among people by using terms like *brother*, *sister*, and *sibling* and recognize less closely related people with different terms, like *uncle*, *cousin*, and *friend*, the scientific naming system uses different terms to indicate different degrees of relationship. For example, you may recognize that rats and mice are closely related to each other and both are less closely related to cats. Thus, at the relationship level that is broader than genus, rats and mice are placed together in the same family, Muridae, whereas cats are assigned to the family Felidae.

The hierarchical scientific naming system consists of nested levels, like progressively larger mixing bowls. Each higher level (bigger bowl) is more inclusive; it embraces more diverse types of organisms. The system looks like this for the brown rat:

Kingdom: Animalia
   Phylum: Chordata
      Class: Mammalia
         Order: Rodentia
            Family: Muridae
               Genus: *Rattus*
                  Species: *norvegicus*

The kingdom Animalia includes all "animals" and excludes plants, fungi, bacteria, and viruses. Most species within the phylum Chordata are vertebrates, or animals with backbones (though there are a few, such as lancelets and sea squirts, that do not have vertebrae but are not related to invertebrates like insects, spiders, and clams). The class Mammalia includes all vertebrates that have hair, feed young with milk, and are warm-blooded, but it excludes other animals with backbones, like birds and amphibians. The order Rodentia includes mice, rats, squirrels, porcupines, beavers, guinea pigs, and hamsters—all have continuously growing upper and lower front teeth that are kept short by use. The family Muridae includes all the Old World rats and mice. The genus *Rattus* contains the closely related rat species. Finally, the binomial *Rattus norvegicus* designates uniquely one species, the brown rat sometimes referred to as the Norwegian sewer rat, and excludes its close relative the black rat.

Some species have been further divided into multiple subspecies. This allows scientists to distinguish between subgroups within one species. Dogs are a good example. *Canis lupus* is the binomial for the entire species that includes gray wolves, domestic dogs, and Australian dingoes, among others. *Canis lupus familiaris* is the scientific name for the subspecies of wolves that includes only domestic dogs. *Canis lupus dingo* only includes Australian dingoes. Though subspecies have distinct different physical characteristics, they can usually interbreed and produce viable offspring. Dingoes typically have larger and longer heads than domestic dogs, but dingoes and dogs can and do breed and produce dingo dogs in Australia. Similarly, there are wolf dogs in the United States. When referring to a particular subspecies, it is important to use the trinomial composed of genus, species, and subspecies names.

Using this scientific naming system, you may begin to develop a mental map of the relationships of organisms. This is an incredibly useful exercise for thinking about how evolution shapes the species and their relationships to one another. For example, at the family level you can see that peach, apple, and pear trees are all closely related and included in the family Rosaceae, and that pine trees (family Pinaceae) and oaks (family Fagaceae) are very different from them! Classifying species is just one of the tools that naturalists and scientists use to conceptualize and communicate about the natural world.

Geographic Range and Landscape Diversity

Each species has tolerance limits beyond which its members are not able to live. These tolerances are dynamic and can change with time and conditions, but at any given time they determine the geographic range of a species, that is, the area in which it can be found. Plant and animal species shift their ranges over time. They are constantly probing the boundaries of their ranges to see if they can be expanded. Geographic ranges also change as climate and topography change. The determining factors that dictate where a species of plant can grow are things like soil, aspect, topography, steepness of and position on the slope, elevation, rainfall, wind, temperature, and distance from the coast. The same is true for many animal species, whose distributions are limited by environmental conditions as well.

Topography, wind, and distance from the coast all play roles in whether a plant is able to grow in a certain place or not. Think of the Coast Ranges of California as a wall blocking the flow of air from the Pacific Ocean eastward. The lowest gap in the Coast Ranges is the Golden Gate, and that is the first place air (wind) pushes through the mountains to reach the Central Valley. This gap allows cool, moist Pacific air to move inland daily, bringing moisture and lowering temperature everywhere it penetrates. Contrast this with the imposing bank of hills along the Big Sur coast, extending from Monterey to Morro Bay. This range of hills (the Santa Lucia Range) effectively stops the flow of cool, moist Pacific air from reaching inland. As a result, at Vallejo, 30 miles inland from the gap between San Francisco and Marin, the temperature at 6 p.m. in July might be 55 degrees and the relative humidity 80 percent, while 30 miles inland from the Big Sur coast the temperature might be 82 degrees and the relative humidity hovering at 17 percent. These are very different conditions for plants and animals to cope with, and consequently there are differences in the species found at these locations.

Aspect generally refers to orientation, or the direction that a slope faces. Different sides of a mountain or hill will have different aspects and will therefore be exposed to the sun for different portions of the day and different parts of the year. Here in the Northern Hemisphere, the sun shines from the south, so south-facing slopes are hotter and drier than north-facing slopes. It is typical in coastal California for cool, moist north-facing slopes to support more conifer trees (for example, redwood, Douglas fir), while hot, dry south-facing slopes are often covered in oaks or chaparral. A similar though less intense distinction can be drawn between east- and west-facing slopes.

Closely related to aspect is slope. Slope is a measure of steepness, with vertical (90 degrees) being one extreme and flat (0 degrees) being the other. The flats at the bases of hills collect soils washed off the hills by landslides and streams, so soils on steeper slopes tend to be shallower and rockier than soils on flats. Plant communities differ depending on where on the slope they are located, from the flats at the base, to the steep midslope, to the gently rounded top of a hill.

Airflow up and down hills varies with the topography and also shapes natural communities. In general, warm air rises, while cold air sinks. Thus, during the day, especially in the afternoon, wind tends to blow up canyons, while cool air tends to flow down canyons at night. This can be quite pronounced, and temperatures at the base of a hill can be very different from those at midslope or at the summit. Cool air tends to pool at the base of hills, causing valleys to be covered in frost or fog while the hills around the valley may be frost-free.

Temperature and rainfall, both daily and seasonal, are huge influences in determining plant and animal distributions. It is not the averages but the extremes that generally determine geographic ranges. Imagine a plant that can tolerate 35 degrees but is killed when the temperature dips below freezing. It doesn't matter if the weather is balmy and warm most of the time. If the mercury dips below freezing one day per year in a specific valley, that plant will be excluded from the site. The same can be imagined on a multiyear scale. Suppose you have a run of 20 "good" years, when the temperature never dips below freezing, but every twentieth year there is a freeze. Our hypothetical plant will still be excluded. We are all familiar with this phenomenon because we hear news reports every 7 or 8 years of the citrus crop in Southern California being wiped out by a cold snap. Even more extreme, every 20 to 50 years we hear about the citrus trees themselves being killed.

Topographical factors such as slope, aspect, altitude, and proximity to water bodies play a determining role in local temperature and rainfall. Higher altitudes tend to be cooler, as air pressure decreases with elevation, allowing the air molecules to spread out, leading to lower temperatures. Mountain ranges often obstruct prevailing winds, causing their windward sides to experience increased rainfall as moist air is forced upward, expands, and cools, resulting in precipitation. Conversely, the leeward sides, experiencing the rain shadow effect, tend to be drier as descending air condenses, warms, and retains its moisture, creating arid conditions. Additionally, valleys and plains nestled between mountains can have microclimates due to trapped warm or cool air that further influences local temperature and precipitation patterns. Death Valley's exceptional dryness results from its location on the leeward side of both the Coast Range and the Sierra Nevada. Moist air pushed over the Sierra produces rainfall on the windward side, but as the air descends on the leeward side, it warms, condenses, and the now-parched air heats up further due to the valley's low elevation, intensifying the scarcity of precipitation in the region. As you may have observed, the plant communities that grow on the windward and leeward sides of the Sierra Nevada are quite different.

GETTING OUT AND ABOUT

Learning about the world never ends. Every year, more plants, insects, birds, and mammals are described for the first time by scientists. These species are "new to science." But an equally interesting category is "new to you." Take time to notice something around you—birdsong, trees planted in your neighborhood park, or native bees visiting flowers. These experiences offer opportunities close at hand for enriching your life and deepening your understanding of the natural world. While outside observing nature, pay special attention to current weather conditions, topography, and other physical factors that may explain in part which species you see and the composition of surrounding plant and animal communities. Conversely, the plants and animals that you observe can tell you a lot about physical conditions such as the climate and soils. Sight, sound, smell, and touch, when safe and appropriate, can all provide valuable clues about the nature of a species. Sometimes even taste can be included in your exploration. Go forth and observe the world, report back to your fellow humans, and engage in Earth stewardship.

Explore!

THE TEN-MINUTE NATURALIST

Spend 10 minutes every day or a half hour once a week outside at the same spot observing. Change the time that you go, if possible, to observe differences between morning and evening. Write down what you see: birds, plants, insects, flowers, and the condition of trees or a creek. Begin the first session by simply walking the spot's perimeter and noting its contours and boundaries and making a simple map of the landscape and important features within it (large trees, structures, streams, fences, etc.). Over time you will deepen your understanding of how nature works. It's best to choose a safe place that's easy to get to, so you can visit it often.

START YOUR OWN NATURE JOURNAL

Nature journals are a time-honored tradition among naturalists and explorers. They can be simple, like a small spiral notebook with quick notes in it, or elaborate affairs with drawings and specimens pressed into them. Whatever kind you choose, remember to include the date, time, location, and weather with each entry.

TRIBES AND TRIBAL COMMUNITIES

Learn the name of the Native Territory you live on and about the local Tribal communities (see https://native-land.ca/). Challenge yourself to learn more about the history of the lands you inhabit and how to actively be part of a better future for Indigenous people, stewardship, and sovereignty moving forward.

2

Geology, Climate, and Soils

Geology happens so slowly, it seems not to move. But don't be fooled. This chapter is about connections between something very solid and real, the Earth we walk on, and something so imaginary it disappears—time. The challenge is to find the connections between the world you see, the rocks you touch, the dirt that grows your food, and the immense spans of time it took to produce them. There is another idea here that bears mentioning. Many of us live our day-to-day lives with the unspoken assumption that the Earth will not change. We imagine that the world of tomorrow will be like the world of today, that no cataclysm will sweep our comfortable existence away. We don't think a flood will wash us downriver, but for the Indigenous people living in the Truckee River canyon during the Pleistocene when the ice dam broke and the top 300 feet of Lake Tahoe came roaring down the canyon, that exact thing happened. Geology teaches that the Earth is ever-changing. Enjoy this chapter, and think about connections between waves and cliff retreat, between earthquakes and tsunamis, between changes in the atmosphere and the advance and retreat of glaciers. Think about the connections of Earth's physical processes with the community of life on Earth that we are all a part of.

As a first approximation, the topographic shape of California can be thought of as a bathtub with a ring-shaped set of mountain ranges enclosing a flat, central valley. The next layer of complexity is that the peaks of the Coast Ranges are lower than those in the Sierra Nevada

Thinking in Geologic Time

We tend to think of the world as always having looked about the way it looks today, but one of the first conceptual abilities a naturalist must develop is the facility to imagine huge expanses of time. When we drive across the Golden Gate Bridge today, we see it crossing the mouth of a huge bay. On a clear day we look 25 miles west to see the granite spires of the Farallon Islands. Ten thousand years ago there would have been no bridge for us to cross, but we would have been able to walk on dry land to the Farallones! If we had taken that walk 35 million years ago, we would have passed through a forest with avocado trees! These lower sea levels also meant people from northeastern Asia could have walked to Alaska when the Bering land bridge was above sea level.

(2,000–8,000 feet vs 11,000–14,000 feet), so the west side of the tub is shorter than the east side. There is one large gap at San Francisco Bay, where the rivers that flow down off of the west slope of the Sierra, and that are gathered in the Central Valley, cut down to sea level through the Coast Ranges to reach the Pacific Ocean.

North of San Francisco Bay, the Coast Ranges form a mostly continuous crest extending all the way to Oregon, while south of San Francisco Bay, they continue uninterrupted to Ventura. The spine of the Sierra Nevada bends west at its southern end and meets the Coast Ranges, and mountains then essentially continue south unbroken all the way to Baja California. These north-south mountain ranges also link up with the Cascades through Oregon and Washington, and on north to Alaska.

At the northern end of the Central Valley, also called the Great Valley, the Coast Ranges are linked to the Sierra-Cascade axis by the Klamath and Siskiyou Mountains. At the southern end of the Central Valley, the Coast Ranges are linked to the Sierra by the Transverse Ranges.

The mountain ranges south of the Transverse Ranges are called the Peninsular Ranges. So, the backbone of California is formed by one tall range of mountains, the Sierra-Cascade axis, which joins the east-west-trending Transverse Ranges, and then the north-south-trending Peninsular Ranges.

This mountain crest running north-south, dividing eastern California from the west, is very important climatically and biologically. To the west of the Sierra-Cascade axis, the land is called cismontane California, and east of the mountains it is called transmontane California.

Relief map of California. A ring of mountains (pinkish brown) surrounds the Central Valley (green). Water appears deep blue. Courtesy of the National Aeronautics and Space Administration.

East of the Sierra-Cascade axis are more north-south-trending mountain ranges. These additional ranges include such long and important features as the Inyo-White Mountains, the Panamint Range (which is the west wall of Death Valley), the Warner Mountains of northeastern California, and a number of smaller ranges scattered across the desert east of the Sierra-Cascade axis.

The final features to plug into a mental map of California are the one bump in the otherwise flat Central Valley (the Sutter Buttes), and the California islands. The Sutter Buttes are the eroded remnants of a

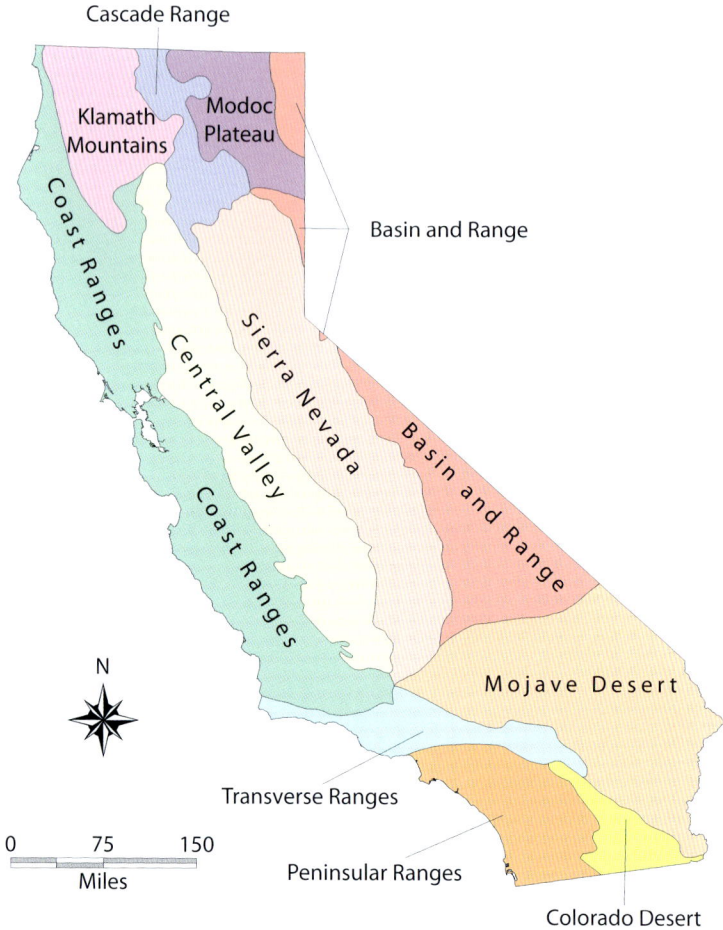

California geomorphic provinces. The mapped geological and topographic features correspond closely to regions distinguished by organisms. Courtesy of the California Geological Survey.

volcano that arose on the floor of the flat Central Valley about 1.5 million years ago (mya) and are perhaps best thought of as the southernmost outlier of the volcanic Cascade Range. The California islands are a series of offshore lands west of Los Angeles, Ventura, and San Francisco. The Southern California islands are called the Channel Islands, while the minuscule dots west of San Francisco are called the Farallones. The Los Angeles basin is a depositional plain enclosed to the north by the

Mobius Arch, Alabama Hills. Photo by Gerald and Buff Corsi.

Transverse Ranges, and to the east by the Peninsular Ranges, making it a part of cismontane California. The Great Basin Desert, the Mojave Desert, and the Colorado Desert are the huge expanses of land east of the Sierra-Cascade mountains and the Peninsular Ranges.

EARTH'S FORMATION AND PLATE TECTONICS

One of the biggest challenges to studying geology is grasping the temporal scale usually involved in geologic processes. For example, the western edge of North America, where California sits, started to take shape approximately 200 to 300 mya. This is when the North American continent collided with the ocean floor, scraping off and crushing oceanic sediments to form mountains that defined the western edge of North America. The evidence of this edge can now be seen as the eastern edge of the Sierra Nevada, with the more recent western part of the state forming from slower sediment accumulation. Ever since this event, more collisions, trenches, and breaks along fault lines have created the landscape we see today.

The topographic structure of Earth's surface is a result of global, regional, and local geological processes that occur over long periods of time. The Earth has a layered structure with a solid inner core, a molten outer core, and a mostly malleable mantle. The upper (outer) mantle

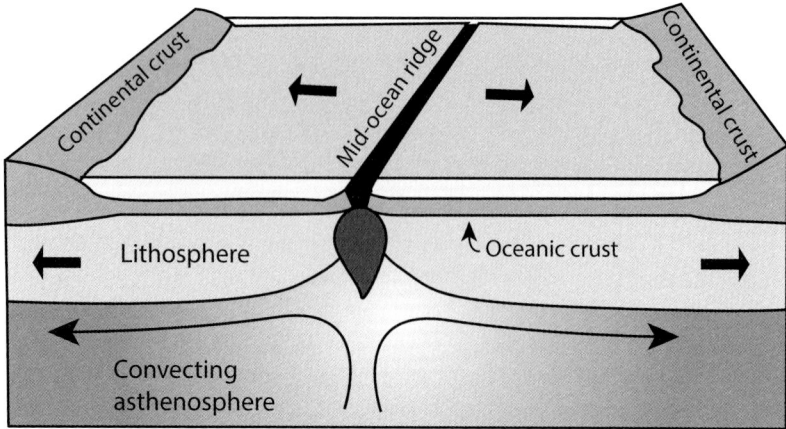

Divergent plate boundary showing relative directions of movement. Courtesy of Rebecca Perlroth, Santa Rosa Junior College.

exists under such extreme pressure and at such high temperature (near the melting point) that although it is solid, it can move like an extremely viscous fluid. Adjacent to the mantle is the outermost layer, called the crust, far thinner than the skin of an apple in comparison, a relatively cool, solid skin for the Earth. It is here that we live our lives. The uppermost part of the mantle together with the adjacent brittle crust is called the lithosphere.

The lithosphere is composed of a series of relatively separate pieces called plates. These are like pieces of a jigsaw puzzle and are pushed about by the motion of the mantle. The plates come into contact with each other, in which case they bump and grind, slide one over the other, rub alongside each other, or crash head-on and crumple at the margins. We call these plate motions and interactions tectonics. Plate tectonics is a relatively new unifying theory, but now that we understand it, geologists can coherently explain earthquakes, volcanoes, and mountain building.

There are fundamentally three different types of plate boundaries: divergent, convergent, and transform. Divergent plate boundaries are locations where plates move apart or separate from one another; at convergent plate boundaries the plates push into each other, and at transform boundaries plates run alongside each other. These different types of plate boundaries are associated with different forces. A divergent plate boundary is driven by tensional forces, a convergent plate

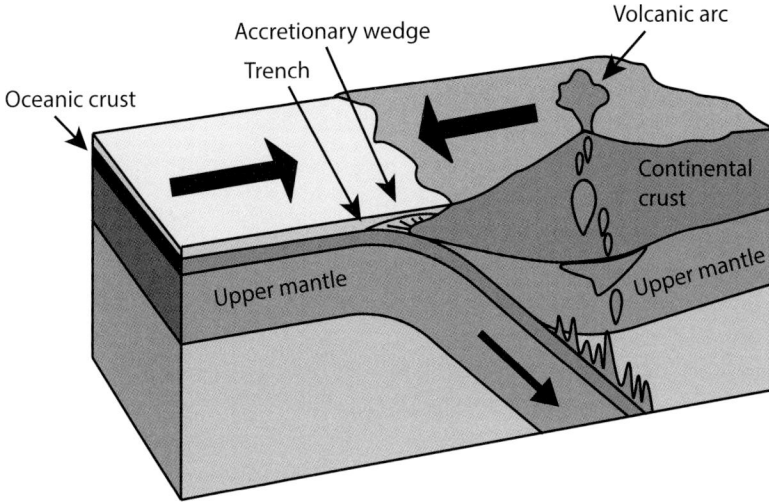

Convergent plate boundary. Oceanic crust usually is subducted beneath continental crust because oceanic crust is more dense. Courtesy of Rebecca Perlroth, Santa Rosa Junior College.

boundary is associated with compressional forces, and a transform plate boundary is driven by shearing forces.

When plates separate at divergent plate boundaries, this allows molten mantle material to squeeze into the gap. The resulting elevated area is called a mid-ocean ridge or spreading center, as it is usually located in the middle of an ocean basin and can contain varied topography, like rift valleys and adjacent ridges. The tensional forces that pull plates apart can result in crustal thinning, which leads to decompression melting and the production of magma. Thus, separating plates don't simply allow preexisting magma to rise up, but rather the motion associated with divergence can be responsible for the production of magma.

Because convection (the process by which material within the Earth's mantle flows in currents) continually repeats itself, ocean basins get progressively wider over time. If plates were all diverging from one another, and new crust were continually being formed, think about what would happen to the Earth. More and more crust on Earth's surface would unavoidably increase the size of the planet! We know that this is not happening, and therefore, if crust is created at one location, it must be recycled back into the mantle in the process of subduction elsewhere.

Surface rupture caused by movement along the San Andreas fault in 1906. Fault trace 2 miles north of Olema, near Point Reyes (Marin County). Tomales Bay in the distance. Courtesy of the US Geological Survey.

Both oceanic and continental lithosphere (plates) can bump into one another. We call this a convergent plate boundary. Because oceanic lithosphere is denser than continental lithosphere, a structure called a subduction zone is formed when oceanic lithosphere collides with continental lithosphere at a convergent plate boundary. The denser oceanic plate sinks (is subducted) into the mantle, where it is eventually reabsorbed (by melting) or rises as magma.

When two continental plates collide at a convergent plate boundary, the result is somewhat different. Because continental crust is so buoyant, it will not subduct into the mantle. This would be like expecting an ice cube to sink to the bottom of a glass of water. Instead, the two continental plates simply push into each other, like two pieces of bread pushing together, creating a huge thickening of the crust. The material cannot go sideways, so it gets pushed up and down. The Himalayas in Asia are a fantastic example of two continental plates colliding. India is moving north and colliding with Asia, causing the Himalayas to rise about 0.4 inches per year or 6.2 miles per million years.

The third type of plate boundary occurs when two plates slide horizontally past one another in opposite directions. This is called a transform plate boundary. The San Andreas fault zone is perhaps the best-known and best-studied transform fault system on Earth. It is highly active, has produced big earthquakes with devastating consequences for Californians, and spans over 800 miles through remote as well as densely populated urban areas. The land west of the San Andreas fault system is part of the Pacific Plate, and the land to the east is part of the North American Plate. These plates slide past one another horizontally, with the Pacific Plate moving relatively to the northwest and the North American Plate moving relatively to the southeast.

SHAPING CALIFORNIA

California is unusual geologically in that all three types of plate boundaries are responsible for the formation of the state and all three are currently active within the state. Divergence is occurring in Southern California, in an area called the Salton Trough, where tensional forces are stretching and thinning the lithosphere, resulting in a large depression in the landscape, the Salton Sink, 269 feet below sea level. The Salton Trough was formed primarily through movement along the San Andreas fault system and the East Pacific Rise. The magma in the trough is shallow and allows for geothermal activity, including hot springs, mud pots, and geysers. Between 20,000 and 3,000 years ago the area contained Lake Cahuilla and remains an inland sea that is separated from the Gulf of California by sediments deposited by the Colorado River. If these sediments were breached, the Salton Sink would again be joined to the Gulf of California, as it has been in the past.

Convergence is occurring north of Shelter Cove (Humboldt County), where the Juan de Fuca Plate is subducting beneath the North American Plate. Convergence along the western margin of the North American continent ended around 30 mya and became lateral slippage to create the San Andreas fault transform boundary that we are familiar with today. Transform plate motion is the primary cause of some of the most notable earthquakes of the twentieth century in California. In 1906 an earthquake along the San Andreas fault caused 21 feet of horizontal displacement and started a devastating fire that destroyed San Francisco. The 1989 Loma Prieta earthquake caused the collapse of the bridge from San Francisco to Oakland, and the 1994 Northridge quake caused the collapse of numerous bridges, highway overpasses, and other structures in the San Fernando Valley area of Los Angeles.

Cascadia quakes are a specific type of earthquake occurring along the Cascadia fault that extends from Cape Mendocino in California to Vancouver Island in Canada. It moves, on average, only every 400 to 600 years. But when it moves, it causes monster quakes that send tsunami waves across the Pacific Ocean that wipe out villages in Japan! In Japan there is a tradition related to Cascadia quakes in California. They erect "tsunami stones" to mark the high-water mark of these rare killer waves so descendants 500 years in the future will know how far uphill they need to run to be safe.

Major parts of the state are west of the San Andreas fault and thus are not, geologically speaking, part of North America. They are riding the Pacific Plate as it glides north. The Farallon Islands off San Francisco actually originated on the Big Sur coast and have rafted north from there! Stick around for millions of years, and Los Angeles will be just west of San Francisco.

One result of all this plate movement is the creation of mountains. The Cascade Range volcanoes of Northern California were formed by the melting and upwelling of subducting plates. The Coast Ranges were formed when the Farallon Plate subducted below the North American Plate and the seafloor crumpled and folded. These mountains are an accretionary wedge, that is, seafloor sediments scraped off the top of the Farallon Plate. The resulting mix of rocks is called the Franciscan complex. Most of the landforms we see in California are only a few million years old, young in geologic time, and mountain building is still in an active phase.

We think of mountain building as a slow, long-term process, and on a human scale it is. But it is important to put actual numbers on some of the geologic phenomena we talk about, because geological events can occur on surprisingly short timescales. Mt. Shasta is the second-highest peak in California and is a classic, cone-shaped volcano (stratovolcano). But how old is Mt. Shasta? Shasta has erupted intermittently for the last 100,000 years. Three of the four major cones that form the current mountain are 12,000 years old or less, with much of the mountain being rebuilt in the last 2,000 years! On a geologic scale, that is surprisingly young. Another important ongoing geological phenomenon is erosion. Shasta, like every other mountain in the world, is constantly being eroded. Landslides, rockfall, glacial action, wind, rain, snow, running water, ocean wave action, freezing, and thawing constantly gnaw away at mountains, transport materials, and fill in low areas, in an apparent attempt to make the whole world completely flat. If it weren't for continuing growth, mountains would (relatively) quickly disappear.

When a mountain like Mt. San Gorgonio is ripped apart by erosion, the resulting debris and sediments end up somewhere below. They pile up, and they often form sedimentary rocks. The flatness of the Los Angeles basin and the Central Valley are the result of this process, where sediments were deposited as rivers ran off the mountains toward the Pacific Ocean over the last 10 or more million years. The sediments in the current Central Valley may be as much as 30,000 feet deep!

Another important geological force that has shaped the landscape of California is the action of glaciers. As global climate has vacillated over the last 3 million years, the Earth has experienced a series of much colder and then warmer phases. During the colder phases, glaciers formed in the higher altitudes of the state and pushed their way down mountain slopes, carving rock and carrying sediment with them. The Yosemite and Hetch Hetchy Valleys are perhaps the most famous glacially carved valleys in the state. The vertical walls of these striking valleys resulted from glaciers cutting through very hard, stable stone, which then stood as huge walls when the glaciers retreated (melted). Since the retreat of the glaciers 15,000 years ago, rockfall has been the main force reshaping the rock walls of Yosemite and other glacially carved faces.

ROCKS IN CALIFORNIA

The movement of plates influences the kinds of rocks found throughout the state. As the Farallon Plate was subducted beneath the North American Plate, there were largely three possible outcomes for the rocks: They could have melted to become magma and returned to the Earth's surface as extrusive igneous rocks (volcanic rocks) or as intrusive igneous rocks (for instance, granite). They could have changed under the extreme pressure and high temperature in the lithosphere and become metamorphic rocks. Or they could have been scraped off and retained on the surface, growing together with the continent to form metasedimentary rocks (both metamorphic and sedimentary rocks).

When a plate is subducted and the leading edge is melted, the magma may work its way up through the overlying continental rock. Magma that reaches the surface hot and semifluid forms what we call extrusive igneous rock. Extrusive igneous rocks are called volcanic rocks. If the rising magma cools to form rock before reaching the surface, it forms intrusive igneous rocks. Granite is the most widespread and familiar intrusive igneous rock in California. Mt. San Gorgonio, the highest peak in Southern California, at 11,503 feet, is composed of intrusive igneous rocks. The important point is that both intrusive igneous rocks (like granite) and extrusive igneous rocks (like obsidian and rhyolite) have similar origins; they come from magma. Either the magma crystallized underground, cooled slowly, and produced large crystals (plutonic rock), or it made its way to the surface, cooled quickly, and solidified with small or no crystals (volcanic rock). In California, Mt. Lassen is a convenient place marker for the dividing line between volcanic rocks found in the Cascade Mountains to the north that were produced by subduction, and volcanic rocks to the south that were produced as a result of the Earth's crust being pulled apart (extensional tectonics), like the Mono-Inyo chain of volcanic craters, Mammoth Mountain, the Long Valley caldera, the Big Pine Volcanic Field, the Coso Range volcanics, and the volcanics of the Salton Sink.

If sedimentary limestone is subducted at a plate boundary and then later is uplifted to the surface again, it may return changed, or metamorphosed. The atoms of the rock rearrange themselves into the smallest, tightest form they can under the influence of extreme pressure and temperature deep in the Earth, so when the limestone returns, it may have changed to what we call marble, a metamorphic rock. Sedimentary sandstone may be changed into metamorphic quartzite, a harder, finer-grained version of the original sandstone. Sedimentary shale can be metamorphosed into slate or schist.

We call the transformation of one rock type into another by geological processes the rock cycle. On a long timescale, rocks are continuously recycled: Oceanic crust may be subducted, melt, and rise through continental crust to become a volcano. The volcano can be worn down by erosion and contribute materials that become a sedimentary rock. The sedimentary rock may be subjected to heat and pressure and metamorphosed. Rock cycle processes are ongoing and occur at a broad range of timescales, from almost instantaneous (like a rockfall or a landslide) to millions of years.

All three of these rock types (igneous, sedimentary, and metamorphic) occur widely in California. Depending on where you are in the Golden State, you may not need to travel far to see all three. Great examples of

Tilted sedimentary rocks. Photo by Gerald and Buff Corsi.

sedimentary rocks can be found all along the lower western slopes of the Sierra, as well as through much of the Coast Ranges. The largest limestone caves in California occur around Lake Shasta. The spine of the Sierra, as well as the crests of the Transverse and Peninsular Ranges, are granitic. The Coast Range between Monterey and Ventura is also granitic.

Common igneous rocks in California are

- Granite and granodiorite: coarse-grained, light-colored intrusive igneous rock. It forms the bulk of the material that makes up the Sierra Nevada, including many of the high peaks, and is prominent in the Transverse and Peninsular Ranges, as well as in parts of the Coast Ranges.
- Basalt: fine-grained, black to rust-colored extrusive igneous rock. Basalt is common in many of the volcanic areas in California, including the Modoc Plateau and the Mojave Desert. One can view dramatic columnar basalt at Devils Postpile National Monument near Mammoth Lakes.
- Other volcanic rocks: rhyolite and andesite form from lavas that have more silica and are stickier than basalt. Tuff is a light-colored, often low-density rock formed when lava is ejected into the atmosphere. Pumice has a frothy texture and is filled with holes

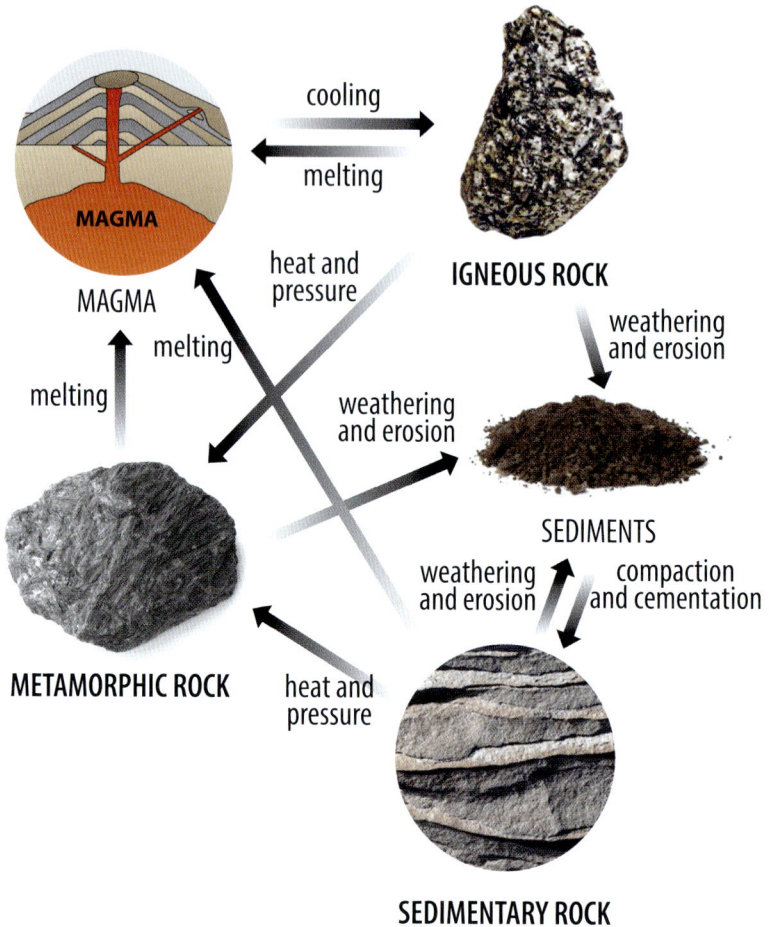

cooling

melting

MAGMA

MAGMA

IGNEOUS ROCK

heat and pressure

melting

weathering and erosion

melting

weathering and erosion

SEDIMENTS

weathering and erosion

compaction and cementation

METAMORPHIC ROCK

heat and pressure

SEDIMENTARY ROCK

The rock cycle. Multiple pathways of change as rocks are subjected to erosion, deposition, heat, and pressure. Used with permission from the National Energy Education Development Project, www.need.org.

making it lightweight. Obsidian is a black and glassy rock formed when lava cools quickly and no crystals are formed. All of these materials are common in many locations with ongoing or past volcanic activity, including the Cascade Range volcanoes and the eastern Sierra volcanics between the Mono Basin and Ridgecrest.

Common sedimentary rocks in California are

- Shale: extremely fine-grained sedimentary rock (formed from lithified mud). Shale forms in calm environments such as deep marine and lacustrine (lake) settings and is common in the Coast Ranges and the foothills of the Sierra Nevada.

- Sandstone: medium-grained sedimentary rock (formed from lithified sand). It forms in continental and nearshore marine environments and is common in the Central Valley and desert provinces but can be found in all provinces of California.

- Chert: organic sedimentary rock composed of layers of microscopic silica-rich shells deposited on the deep marine floor. Chert is generally rust red in California but can be deep green, yellow, orange, or gray to black. It is common in the Coast Ranges and in the foothills of the Sierra Nevada.

- Limestone: a sedimentary rock composed of layers of carbonate-rich shells deposited in generally shallow marine waters, such as reefs. Limestone is prominent in some of the desert mountains of the eastern Mojave.

Common metamorphic rocks in California are

- Serpentinite: the state rock of California! This rock is green to black, waxy, and extremely soft. It is the result of tectonic compression and hydrothermal alteration of dark igneous rock (like peridotite and basalt) at an ocean-continent convergent boundary. Serpentinite is common in the Coast Ranges and the foothills of the Sierra Nevada.

- Slate: fine-grained foliated metamorphic rock produced from alteration of shale or mudstone. Slate is common in the foothills of the Sierra Nevada.

- Schist: medium-grained, sparkling, foliated metamorphic rock produced from alteration of sedimentary or volcanic rocks. Schist is common in the Coast Ranges, the foothills of the Sierra Nevada, and the Transverse Ranges. It is common to see a particular variety of schist, called blueschist, in the Coast Ranges.

CALIFORNIA'S CLIMATE

The climate of lower-elevation California west of the Sierra-Cascade axis is entirely different from the climate found across the rest of the North American continent. It is so different from the continental climate east of the Sierra-Cascade axis, it might as well be on a separate continent, or its own island. Elna Bakker titled her wonderful introduction to California biodiversity *An Island Called California*, in homage to this contrast. In California, continental climate is turned completely on its head. Instead of the wet season and the warm season occurring at the same time, they are mostly nonoverlapping. The warm season (May to October) in lowland California is characterized by unrelenting drought. It virtually never rains in most of California during the warm season (exceptions are summer thunderstorms over the high mountains and coastal fog). On the other hand, during the "cold season" we get most of our rain. However, it seldom freezes at lower elevations in California. Rain falls in California during the cool season, from November to April. This reversal of conjunctions—warm with dry and cool with wet—and the lack of extreme (freezing) cold are the prime characteristics of a strange, globally rare climate type.

The common term for this climate type is Mediterranean climate, and it is only found in four other areas in the world, all on the western edges of continents, roughly between 30 and 40 degrees latitude: the Mediterranean basin, Chile, southwestern Australia, and the Cape region of South Africa. These five areas have more in common with each other climatically than any do with the geographically adjacent areas of their own continents. Climatically, California west of the large eastern mountain ranges is more like Greece or Chile than it is like Nevada. To the extent that climate determines the direction of evolution, these five areas have been subject to evolutionary pressures of a very similar type.

Given that Mediterranean climates are found outside of the Mediterranean basin, maybe we should come up with another name for this type of climate. Any ideas? Another terminology challenge is that with hot, dry summers and mild, rainy "winters," the vocabulary we use to describe our seasons is strangely at odds with California organisms. If "spring" is the time of year when seeds sprout and perennial herbs initiate growth, then spring in California is in November, not April as in most of the continent! Spring in November? Yes! If "winter" is when mammals hibernate, that would be November to March in Siberia but would be June to October in California. We even have a separate word for going dormant in the hot, dry season. We say that animals estivate rather than saying that they hibernate. If summer is the prime growing season for plants, when grass grows, summer is May to September in New York. But in California, by this definition, summer is November to

March! California is so strange climatically, so reversed and mixed up, that most of the trees and shrubs fail to lose their leaves at all. In fact, all of the Mediterranean climate areas of the world are characterized by a very high proportion of evergreen trees and shrubs, compared with plants in continental climate areas. These generalizations apply best to low-elevation California (below 3,000 feet) west of the Cascades, the Sierra, and the Peninsular Ranges and are contradicted along the immediate coast and high in the mountains.

DIVERSITY OF CALIFORNIA MICROCLIMATES

The five Mediterranean climate areas of the globe are all on the western edges of continents. Global winds form as a result of air in the tropics heating and rising, flowing toward the poles, then descending as it cools. The spinning of the Earth in one direction causes these global air movements to bend (the Coriolis effect). The result in California is that the large-scale winds usually blow from west to east, from the ocean onto the continent.

In California, this means that the wind generally blows off the Pacific Ocean, blows east across the state and over the mountains, and continues east across the continent. The wind blowing off the Pacific is full of moisture from passing over open ocean. When moist air moves inland and runs into mountains, it rises, and this causes it to expand, which results in cooling. As air cools, it becomes more saturated, increasing the chances for cloud production. The more clouds form, the greater is the likelihood of precipitation. As the moist air mass cools, it drops its moisture, which hits the earth as rain or snow. As clouds rise to pass over the north-south mountain ranges of California, they are wrung of moisture. For each 1,000 feet of elevation gain an air mass experiences, its temperature drops about 3.5 degrees (more when dry, less when wet). As a cloud passes over the 2,000-to-4,000-foot crest of the Coast Ranges, its air temperature will drop 7 to 14 degrees. This drop in temperature causes the air mass to release moisture as rain or sometimes snow on the Coast Ranges.

After passing over the Coast Ranges, the clouds descend into the basin of the Central Valley, increasing in temperature as they descend. The increase in temperature causes the clouds to have additional ability to hold on to moisture and not release it. For this reason, the Central Valley tends to be drier, with lower rainfall than the Coast Ranges. The same relationship exists east and west of the Peninsular Ranges. The west slope of the Peninsular Ranges is well watered; the east side is drier. As the clouds rise to ascend the Sierra and Peninsular Ranges, they expand and cool once again. The mid-elevation forests of the Sierra and Peninsular Ranges get abundant moisture as the clouds rise through elevations they have not previously seen in the Coast Ranges, and the highest elevations of the mountains receive the highest amounts of precipitation.

After being lifted over the crests of the high Sierra Nevada and Peninsular Ranges, air will descend down the steep east sides to lower elevations. In the process of descending, the air will contract and heat up and consequently be drier when it reaches the transmontane basins, where there is little chance of their dropping rain. Consequently, the basins east of the mountains are largely desert country: the Great Basin, Mojave, and Colorado Deserts.

To summarize, the west sides of mountains get more rain; the east sides receive less. The entire Central Valley is in the rain shadow of the Coast Ranges, and the desert lands east of the Sierra and Peninsular Ranges are dry because the rain has been squeezed out of the clouds as they pass over the mountains.

Another pattern prevails as you travel from south to north through the state. Broadly, rainfall increases as you proceed north. In the Coast Ranges rainfall is greater in the Santa Cruz Mountains (north of Monterey) than it is in the Gabilan Range (south of Monterey). Continuing north, rainfall is greater around Mendocino than in Santa Cruz, and still greater in Arcata than in Mendocino, given similar elevations. The same pattern prevails in the Sierra and Peninsular Ranges: the farther north, the more moisture is delivered as rain or snow at the same elevation.

A notable exception to these patterns is the occurrence of summer rain over the Southern California deserts. In summer moist air masses may come north from the Gulf of California, and even come across Arizona from the Gulf of Mexico, bringing rare summer rainfall events to the desert mountains of California. There are annual plants that bloom in August and September when these desert rainstorms prompt them.

The source of California's "wintertime" precipitation is mid-latitude cyclonic storms originating from the Gulf of Alaska and controlled by the jet stream. Since the jet stream typically flows from west to east, and farther north, it less often sweeps south into California than, say, Washington state. It is even less frequent that it sweeps as far south as Central California. And even more infrequently, it moves wintertime storms all the way into Southern California. Hence, locations farther north receive

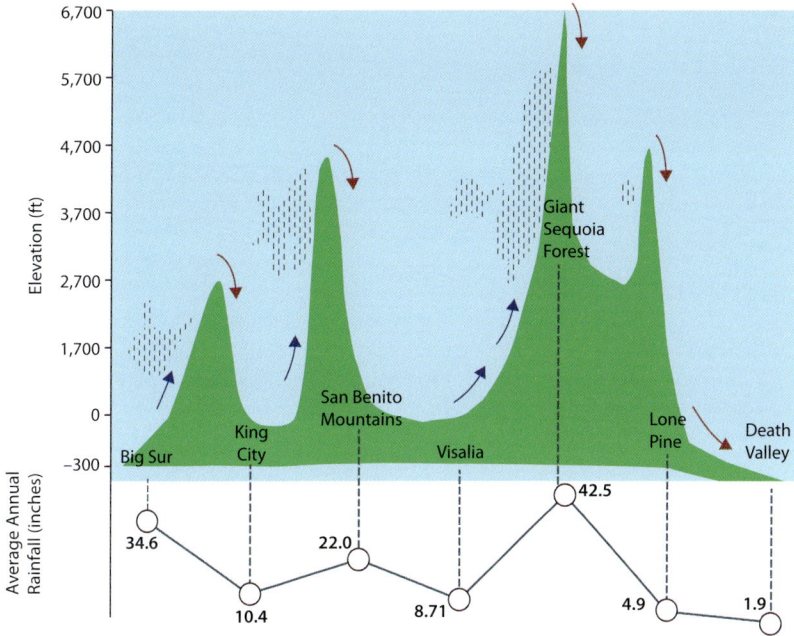

The orographic effect over a cross section of California from west to east and the influence on precipitation patterns, referencing specific locations spanning the Big Sur coast across to Death Valley with average rainfall (inches/year) recorded at these locations. Courtesy of Adina Merenlender.

more precipitation—because they are likely to receive the jet-stream-driven storms from the north more frequently.

The result of this complex interplay of wind, cloud, rain, and topography is a wide range of microclimates. Microclimates are small zones where climatic factors differ from the surrounding area due to localized influences such as slope, aspect, topography, and elevation. These microclimates play a strong role in limiting or allowing the growth of plants and animals on the landscape, since plant and animal distributions are strongly constrained by temperature and moisture.

Not every year is the same in California. Some years there is abundant rainfall; others are very dry. One of the factors that influence the variation in multiannual rainfall patterns is ocean surface temperature in the tropical eastern Pacific. Changes in sea surface temperature influence global atmospheric circulation, which, in turn, shifts the position of the jet stream either southward or northward. These sea surface temperature changes exhibit a somewhat predictable pattern, typically

The California Current

Winds along the California coast typically blow strongly from north to south. Similarly, a strong current called the California Current flows in the ocean just off the California coast, running from north to south. These winds and the coastal topography bring nutrient-rich waters from the deep ocean to the surface along the California coastline. This is called upwelling and results in some of the most biologically productive marine zones on the planet.

The California Current is one of only five such zones in the world. It begins in the North Pacific and travels south along the West Coast until it dissipates near Baja California. The five ocean upwelling zones, which represent only 5 percent of the total ocean area, produce 25 percent of the world's seafood and support creatures large and small, from the blue whale (*Balaenoptera musculus*)—the largest animal ever known to have lived on Earth—to small crustaceans called krill. Upwelling strengthens as winter wanes, reaching its peak in spring and summer. Recent models of climate change predict that upwelling events will become more intense in the spring and less intense in the summer, particularly in the northern part of the California Current, due both to changes in atmospheric pressure and the overall impact of climate change on ocean temperatures. The extent of upwelling is influenced by ENSO and has major implications for the health and abundance of marine wildlife in the California Current.

Adapted from a UC California Naturalist regional publication: *Natural History of the California Current.*

occurring over a period of 2 to 7 years. When the eastern tropical Pacific experiences a warming of 1.8 to 5.4 degrees (1 to 3 degrees C), it impacts both atmospheric circulation and the upwelling process along the California coast. Since upwelling brings nutrients that support fish, ocean fishers in South America first experience these changes. Because the warm water often is most noticeable around Christmastime, this phenomenon is called El Niño (The Child). El Niño events bring intense rainstorms to Southern California. When sea surface temperatures are cooler than "normal" in the tropical eastern Pacific, we now refer to this phenomenon as La Niña (The Girl). This somewhat regular pattern of changing ocean temperatures, jet stream movement, and variations in upwelling is referred to as the El Niño Southern Oscillation (ENSO). Periods when the sea surface temperatures in the tropical Pacific Ocean are near average are referred to as ENSO-neutral—not too hot and not too cold, just right for Goldilocks and her fish friends.

If you are going to remember only one thing about climate in California, it should be that there are many climates in California and that variation from year to year, south to north, east to west, and low elevation to high elevation is legion. But the entire state experiences dry warm summers and wet cooler winters.

SOIL STRUCTURE AND NUTRIENTS

Plants depend on the sun for their energy but must get water and the mineral nutrients necessary for growth from the soil. Soil is formed from the weathering of bedrock and contains both inorganic and organic components. Differing amounts of mineral material, organic matter, water, and air can be found in local soil. Animals and plants that live in the soil affect soil properties, such as the amount of organic matter and water content, and vice versa—vital minerals in soil are absorbed by plants and ingested by animals. Soil scientists classify layers of soil into different horizons according to their characteristics, which are related to depth:

O horizon—surface: organic material, dead plants, animal material

A horizon—topsoil: plant roots, bacteria, fungi, small animals

B horizon—subsoil: fewer organisms, less topsoil; poorer conditions for plant growth

C horizon—altered parent material: weathered, less living matter; materials that formed the layers above

Different rock types have different chemical compositions, and as rocks break down to form soils, the parent rock determines the chemistry of the soil formed upon it. Plants are sensitive to different soil chemistries, and some soils generally promote the growth of all plants, while other soils are inhibiting or even toxic to some or all plants. A dramatic example is the presence of salt in a soil. When rocks are eroded by moving water, salts will be moved by rivers to the sea. But if the rivers or streams flow into an enclosed basin with no outlet to the sea, like Mono Lake and Owens Lake, the salt accumulates in the basin, and a salt lake or alkali lake results. Soils impregnated with high concentrations of salt are toxic to most plants and thus exclude them. Dozens of these dry salt pans exist in the California desert and produce remarkable landscapes,

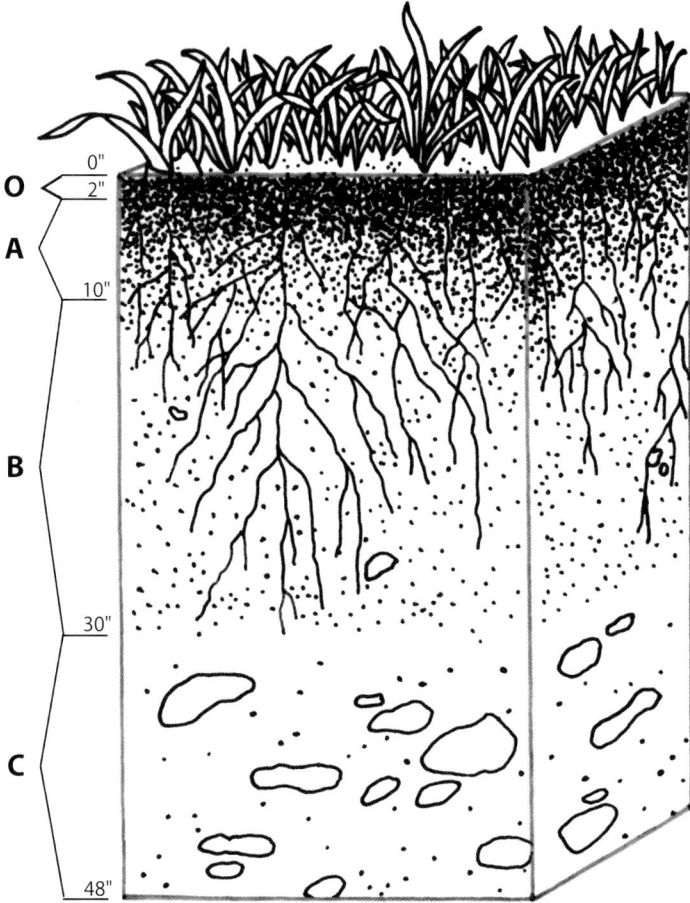

An idealized soil profile. The deeper one digs, the less organic matter and the coarser the rocks and stones. Adapted from a Natural Resources Conservation Service USDA soil drawing.

such as those in Death Valley. If you want to taste the salt, try some pickleweed (*Salicornia* sp) in your salad. Along with salt grass and some rushes, pickleweed can be found in saline environments, such as salt marshes and estuaries, or near springs in coastal regions.

Another unusual type of soil that supports many endemic plants is serpentine soil. Serpentine is a rock associated with subduction zones, crumpling, and accretion at plate margins. Serpentine soils are characterized by a striking calcium-magnesium imbalance. Calcium is an essential plant nutrient. Most soils have lots of calcium and little magnesium, so plants generally ignore calcium. It is a common, easily available plant nutrient. However, serpentine soils have very little calcium but are high in magnesium. Calcium and magnesium are chemically very similar and therefore seem alike to plants. In serpentine soils, generalist plants that don't bother to distinguish between calcium and magnesium take up both indiscriminately, and they are poisoned by an overabundance of magnesium before they can get enough calcium to meet their needs. Thus, many generalist plants and most invasive grasses are excluded from serpentine soils, leaving an open niche that some native California plants have taken advantage of as they have become serpentine specialists. One of these serpentine specialists is the leather oak (*Quercus durata*), which grows only on serpentine. The boundary between serpentine and nonserpentine soils can be very sharp, and the plants growing on it may distinguish the soil boundary visually for observant naturalists.

Another of the essential nutrients plants must acquire to grow is nitrogen. Nitrogen is present in most soils but not in amounts that allow luxuriant growth. Nitrogen and phosphorus are two of the most common "fertilizers" farmers apply to crops to increase growth. Some plants have found a way to acquire nitrogen from the air, thus avoiding dependence on whatever nitrogen may be available in the soil. These plants produce tiny hollow places on their roots (nodules) that are a perfect habitat for specialist bacteria that convert atmospheric nitrogen into a form that plants can eat.

As the bacteria complete their life span of a few hours or days and die, they decompose in the nodule and release nitrogen to the plant, which "fertilizes" the plant. In chaparral, buckbrush (*Ceanothus*) develops nitrogen-harboring nodules. Along streams, alders (*Alnus*) do the same thing, and in grasslands, lupines (*Lupinus*) make root nodules for bacteria. As the alders, buckbrush, and lupines complete their life spans of 1 to 60 years, they decompose and release the nitrogen they have harvested from the atmosphere (via their bacterial symbionts) to the soil, thus fertilizing the soil for other plants. One unusual consequence of the nitrogen abundance provided to alders (a commonly found riparian tree) by their guest bacteria is that alders do not need to expend energy to extract nitrogen from their leaves before shedding them in the fall. Alder leaves remain green to the day they drop!

NUTRIENT CYCLING

Outside of a few infalling meteors and outflying spaceships, the Earth is a closed chemical system.

—William H. Schlesinger

The Earth is an elegant system. Soil, air, and water characteristics change and adapt according to the flow of chemicals, nutrients, and materials through integrated natural cycles. Materials cycling through the Earth's systems go in and out of biotic and abiotic stages. Living things grow, reproduce, die, decompose, and become part of other living things. Nonliving things accumulate, break down, are utilized by or stored in living organisms, and get excreted and redeposited, reformed, or reused. The movement of matter through Earth's systems is described as biogeochemical cycles. Some parts of these cycles take relatively little time, while others can take place over millennia. Among the best-known and most important cycles are the rock cycle, the water cycle, the carbon cycle, the nitrogen cycle, and the phosphorous cycle.

Throughout these cycles, nutrients fluctuate between being available and unavailable for utilization by living organisms. Nutrients are released and become available through decomposition, respiration, excretion, erosion, and combustion. They are taken up by organisms through photosynthesis, ingestion, and assimilation. They are deposited through sedimentation and fossilization. Nutrients remain in these different states, or reservoirs, for different residence times. For example, carbon remains in plants for 5 years on average. By far most carbon on Earth is found in the form of rocks and may remain locked there for millions of years.

Awareness of cycles is critical to understanding Earth processes, as our actions can greatly influence biogeochemical cycles. As part of Earth's systems, we are impacting these cycles, especially through the introduction of materials at a much faster rate than natural cycling processes can absorb them. In fact, scientists are increasingly referring to the current epoch as the Anthropocene, due to the significant impact of human activities on the Earth's geology and ecosystems.

The nitrogen cycle provides an excellent illustration of this point. Nitrogen is important for agriculture because it is often a limiting factor for plant growth. For that reason, farmers often apply large amounts of nitrogen in the form of nitrate to increase crop yields. Unfortunately, modern agriculture applies more nitrogen than plants can absorb. Nitrates are highly water-soluble. What is not used by plants can move out of soils

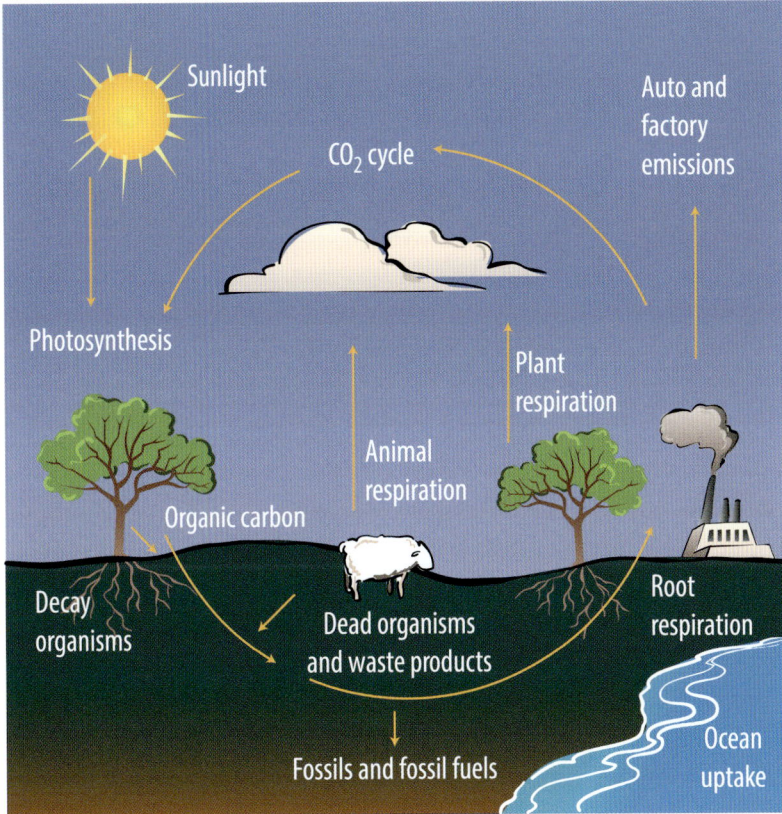

The carbon cycle. Carbon moves via multiple pathways between the atmosphere, the geosphere, living and dead organisms, and the ocean. © University Corporation for Atmospheric Research.

and become a source of water pollution in streams, lakes, and even the ocean. Other land use activities such as mining, timber harvest, and urban development can alter nutrient cycling over long periods of time.

MINING IN CALIFORNIA

One of the by-products of the geologic diversity of California is deposits of ores that are useful to human society. California's love affair with mining didn't end with the Gold Rush. Many people are surprised to learn that a substantial amount of mining continues to this day and is valued at over $5.6 billion per year. Mining products include boron, Portland cement, gold, sand and gravel, bentonite clay, pumice, salt,

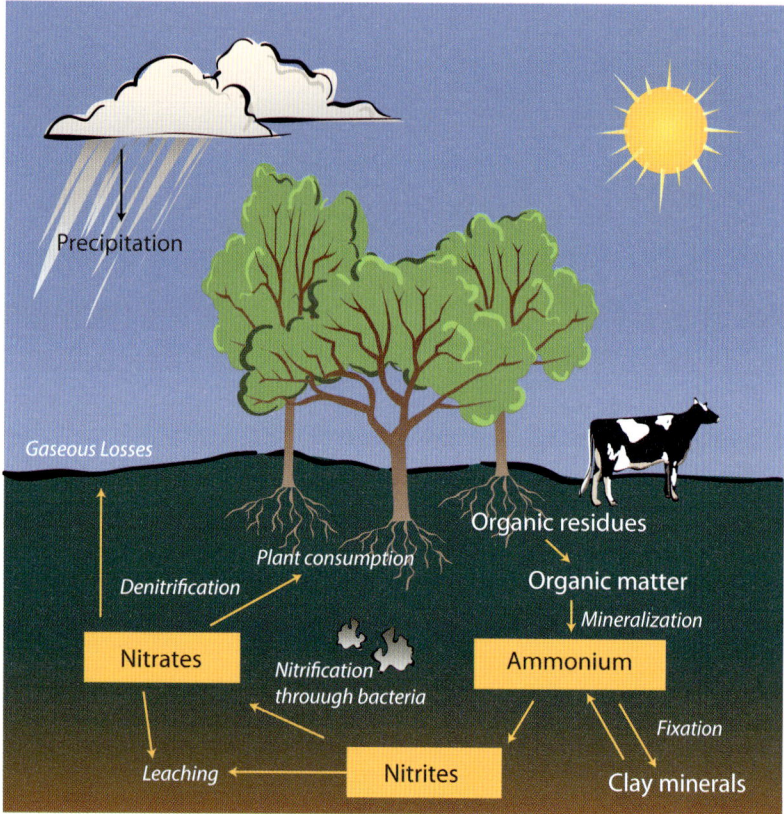

The nitrogen cycle. The nitrogen cycle describes the process by which nitrogen circulates among air, soil, water, and organisms in various chemical forms. © University Corporation for Atmospheric Research.

lime, feldspar, limestone, and kaolin clay, among many others. These are all important industrial materials that modern society depends on. Nonetheless, the impacts of both current and historic mining on the environment can be substantial, including long-term effects on the shape of a river by widening and deepening the channel both upstream and downstream of a mining operation. For example, in-stream gravel mining on the Russian River has changed the shape of the channel, deeply incising it and altering its hydrology, which has impacted the salmonids and other species in this large river.

Many historic mines have environmental impacts that continue to be felt today. Mercury is used in gold mining to separate the gold from the

Naturalists getting out to explore Bishop Tuff. Photo by Judd Curran, Professor of Geography, Grossmont College.

ore. Unfortunately, mercury-laden sediments remain from old gold mines and provide a continual source of contamination of fish, wildlife, and water. One example of long-term mining impacts is the Iron Mountain Mine, a Superfund site near Redding. The Iron Mountain Mine was at one time the largest copper mine in California. Though mining there ended in 1963, an EPA case study of the site states, "Nearly 100 years of mining activity at Iron Mountain left numerous waste rock and tailings piles, massive fracturing of the bedrock overlying the extensive underground mine workings and remaining sulfide deposits, sinkholes, seeps, and contaminated sediments in nearby water bodies. The uncontrolled acid mine drainage from Iron Mountain Mine was the largest source of surface water pollution in the U.S." Through cleanup efforts, the site is now 95 percent contained.

We encourage you to explore the intricate connections between geology and the community of life on Earth and consider the vast spans of time it took to form California as we know it today. We have plate tectonics, along with faulting, folding, erosion, and volcanic activity, to thank for sculpting California's landscape into a big bathtub with the

Central Valley in the middle surrounded by mountain ranges that greatly influence rainfall patterns. Notable geological events such as earthquakes remind us that change can also happen quickly, but mostly what you see comes from slower processes such as erosion and glaciers. The geological diversity of California has resulted in a rich variety of rock types throughout the state. Igneous rocks, formed from magma, include intrusive types like granite and extrusive types like obsidian and rhyolite. Sedimentary rocks, such as sandstone and limestone, are prevalent, but metamorphic rocks, like marble and quartzite that transformed under extreme pressure and temperature, can also be found. California's Mediterranean climate, characterized by dry summers and wet winters, is influenced by its geography, and it means plants are mostly dormant in the summer and come to life when it starts to rain— California springtime. Microclimates vary widely due to topography, elevation, and proximity to coastal or mountainous regions. Soil type and nutrients, or the lack thereof, play a crucial role in determining what plants can be found where. Human activities such as agriculture, urbanization, and pollution influence nutrient cycles, sometimes on a large spatial and temporal scale, and both historic and ongoing mining operations have had lasting environmental impacts on ecosystems.

Explore!

TAKE AN EARTHQUAKE WALK

Point Reyes National Seashore has a popular trail called the Earthquake Walk that follows the fault line. USGS has an excellent guide titled *Where's the San Andreas Fault? A Guidebook to Tracing the Fault on Public Lands in the San Francisco Bay Area* with information about earthquake touring from the Hollister and San Juan Bautista areas to the Santa Cruz Mountains and the San Mateo coast.

TAKE THE CALIFORNIA GEOTOUR

The California Department of Conservation has information about the California Geotour, a wonderful list of geologic field trips throughout the state.

GO SEE A GEOLOGY AND MINERALS COLLECTION

California Academy of Sciences in San Francisco has over 400 specimens to look at. The Los Angeles Museum of Natural History showcases over 2,000 minerals, rocks, meteorites, and gems from across the globe.

GO SPELUNKING

Mercer Caverns, Moaning Caverns, Lake Shasta Caverns, Crystal Cave, Black Chasm Cavern, Mitchell Caverns, and California Caverns are all worth seeing.

3

Water and Watersheds

Water is the basis of life on Earth. It is a primary shaper of our climate and a dominant force in geology. Water connects us to our food, to the atmosphere, and to all the inhabitants of the planet. Something as simple as taking a shower connects most Californians to distant parts of the state, through pipelines and aqueducts. In this chapter, the topic of water provides connections in ways that may surprise you and, hopefully, that will deepen your appreciation of the importance of water to our state.

The physical landscape of California has been shaped by water as a liquid and as solid ice, while the social and political landscape of modern California has been shaped by water management, conveyance systems, and a complex web of water laws that influences the geographical layout of cities and determines who has access and control over this most basic need of all living things.

While water plays an important role in daily political, ecological, and cultural life everywhere, there are some things about water in California that are distinctive: annual cycles of wet and dry periods that are characteristic of the Mediterranean climate, extreme variations in rainfall from year to year, a heavy reliance on spring snowmelt and groundwater, geographic variation in precipitation within the state, and heavily engineered water delivery. None of these are unique to California in and of themselves, but the combination of all at once is inimitably Californian. Understanding these forces helps to make sense of California water ecology and politics.

SCALING WATER: FROM MOLECULES TO OUR ENVIRONMENT

Two hydrogen atoms, one oxygen atom—on its face, water seems so simple. Yet its molecular shape, and the capacity of hydrogen and oxygen to act independently and together, allows water molecules to interact in ways that other molecules cannot. Water is unique at a molecular scale because it can take a solid particle, break it apart, and hold it in a liquid form. The polarity of water (its nonlinear, asymmetrical shape and charge) allows it to dissolve salts and other compounds, such that those compounds can then be absorbed by organisms or transported downstream.

Water comes in three forms—liquid, solid, and gas—and the chemical properties of water that enable these three states have important implications for life on Earth. One of the most important of these properties is that weak hydrogen bonds are regularly formed between adjacent water molecules. These weak bonds allow water molecules to form liquid water, water vapor, and solid ice.

In liquid water, hydrogen bonds form and fall apart quickly and locally (at the molecular level) but provide sufficient structure to give water its surface tension. Surface tension due to hydrogen bonds provides the resistance that allows water striders to skate on the water's surface. In its solid form (ice), the hydrogen bonds are more stable and create an organized lattice of molecules that are less tightly packed, compared with water in the liquid form. Liquid water is thus more dense than ice, resulting in the capacity for ice to float. If ice sank to the bottom of a lake, the lake would freeze bottom to top and entomb organisms living in the lake. Instead, ice formed on lakes, rivers, and oceans remains floating on the top, while the water below, and many of the organisms within it, generally remain above freezing for the entire winter. This unusual physics allows some aquatic plants and animals to survive in freezing conditions.

THE WATER CYCLE

Water on Earth is finite: it is neither created nor destroyed. The water we drink is the same water that was drunk during the time of the dinosaurs, the same water that first appeared on Earth approximately 4.5 billion years ago. So how does water move and change?

If you've ever sat on the bank of a river, it may have crossed your mind that water is constantly flowing from the mountains to the sea. But where does the water in rivers come from? The key to the perpetual

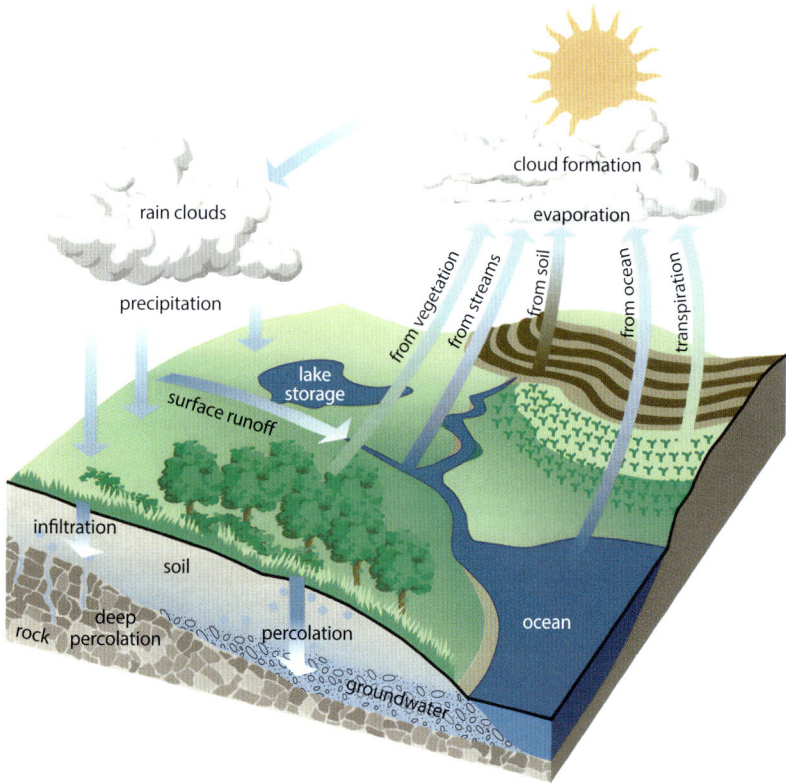

The water cycle. The movement of water from precipitation to surface water and groundwater, to storage and runoff, and eventually back to the atmosphere is an ongoing cycle. Adapted from an image from the Federal Interagency Stream Restoration Working Group, Stream Corridor Restoration, 1998, 2001.

movement of water from sea to air to land and back to the sea is the repeated transformations between its solid, liquid, and gas forms, transformations described by the water cycle.

The water cycle is controlled by the sun and by gravity. The sun heats water, causing it to change from a liquid to a gas, creating water vapor, which moves in the atmosphere, then changes to liquid and falls as rain onto the oceans and the land. Some of the water runs into creeks and rivers, flowing from higher elevation to low; some seeps into air spaces in the soil and creates reserves called groundwater; and some gets locked up for millennia in glaciers or polar ice caps.

About 70 percent of the surface of Earth is covered in water, but fresh water represents only 1 percent of all the planet's water. Think about that: all the lakes, rivers, and groundwater, plus the water present in living things, is only 1 percent of Earth's water. Approximately 2 percent of Earth's water is frozen as glaciers and polar ice caps. Most of Earth's water (97 percent) is the salt water of the ocean.

Like other Mediterranean climate areas, California has annual cycles of wet winters and dry summers. In most years, the majority of rain occurs during a relatively small number of winter storms, the most intense of which are called atmospheric rivers. How frequently and where rain falls is highly variable in California and is dependent on the state's geography, topography, and position on the western edge of the continent.

California is about 800 miles in length from Oregon to Mexico, with a north-to-south rainfall gradation; that is, the majority of rain falls in the northern part of the state, and annual rainfall generally decreases farther south. This difference can be extreme, ranging from an average of 10 inches per year in San Diego to 93 inches on the Smith River in far northwestern California. Moreover, California is subject to wide variations in rainfall from year to year, sometimes referred to as weather whiplash. Also, the state of California can undergo very long periods of drought, lasting years or decades.

Most rain that falls on land is absorbed directly into the ground. Lakes and rivers themselves cover only a tiny fraction of the land surface, and most water that reaches streams does so by traveling in cracks and air spaces underground. When it rains, water soaks into the ground, passes through the soil, and settles into the porous subsurface areas. As these underground features become saturated, the water table rises. When the underground reservoir fills to the point of an exit on the Earth's surface, we may find a spring on a hillside, or the water may move into a stream or river. Approximately 60 percent of precipitation ends up in streams in a typical year. Some of the remaining water remains in the soil and recharges the groundwater reservoir, and a tiny amount is taken up by plants and animals.

Most rainfall reaches streams within hours, so elevated streamflow tends to occur on the same day as a rainfall event. The highest levels of streamflow occur during the rainy season (November to April), when precipitation is greatest in California. If there is a stream near you, you should visit it often, especially during and after storms (if safe), to see how the stream level changes. When you visit during the dry season, look for flood debris: twigs, piles of grass, and small and large wood left at the high-water mark by storm surges. You can also observe remotely

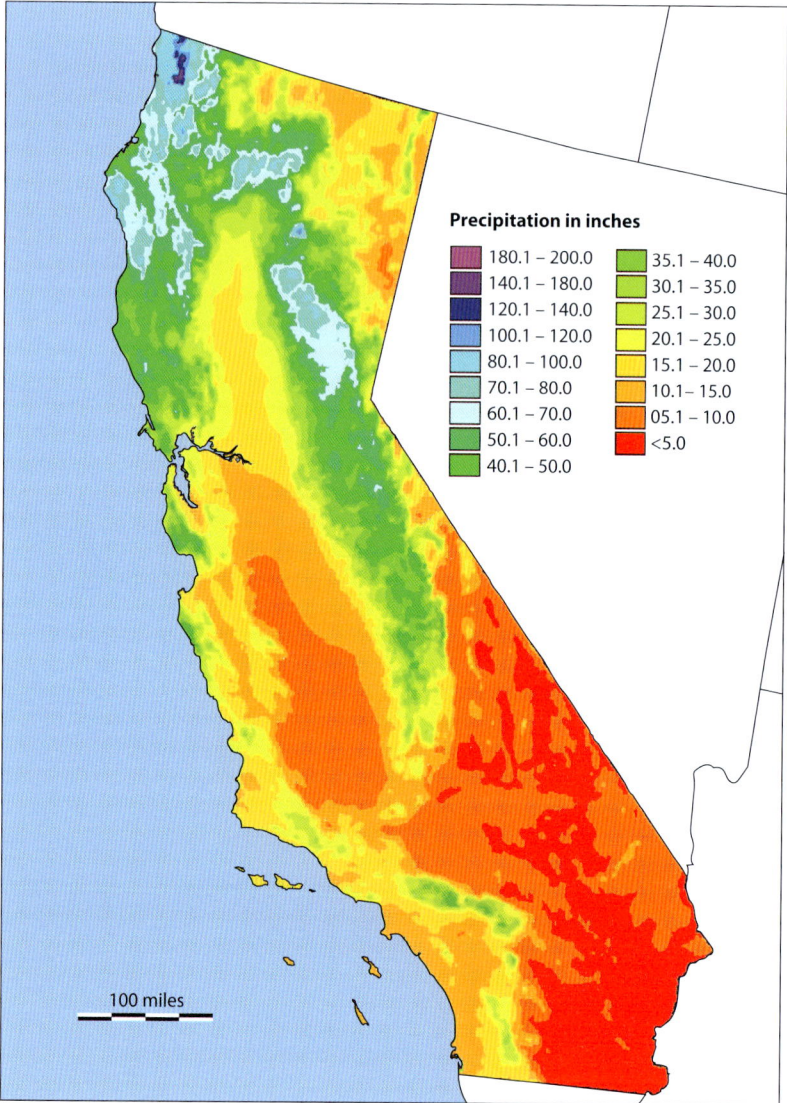

Average annual precipitation in California. Annual average rainfall generally increases from south to north and decreases on the east side of mountain ranges (in rain shadows). Modified from the *National Atlas*, 2005, US Geological Survey, www.nationalatlas.gov.

The Whitewater River, within the boundaries of the Whitewater River Area of Critical Environmental Concern, San Gorgonio Wilderness, Sand to Snow National Monument, and part of the Whitewater Wild and Scenic River. Photo by Colin Barrows.

with data. The USGS operates stream gauges on most California rivers and creeks, and the data can be found at the USGS Water Data website.

Whether precipitation falls as rain or snow has important implications for the timing of the movement of water. Snowfall is retained through the winter and is not converted to liquid water until snowmelt. Snowmelt arrives in streams in late spring and summer when air temperatures in the mountains are sufficiently high. Many ecological processes in the state are dependent on this timing. For example, spring- or summer-run Chinook salmon and trout depend on cold water and suitable flows in late spring and early summer to spawn, rear, and migrate. Humans also rely on spring snowmelt for augmenting water supply, such as filling reservoirs for summer release.

Historically, snow has been a significant long-term repository for water in California's hydrologic cycle. However, with rising temperatures due to climate change, snowmelt is occurring earlier or snow is failing to fall at all. When more of the winter precipitation falls as rain rather than snow, less is available in late spring and summer for human uses, for enhancing streamflows, and for recharge of groundwater. This pattern is exacerbated by an increased frequency of "rain on snow" events, which occur when air temperatures are warm enough for precipitation to fall as rain at high elevations where snow is already on the ground. Snow covering the land acts as a barrier to absorption. The rain melts the snow, causing an effective doubling of effective precipitation; the rapid conversion of rainfall and snowmelt into streamflow tends to cause flooding both locally and in downstream towns.

Most streams in California eventually run into the ocean. East of the Sierra-Cascade axis, however, in the California deserts, many rivers and streams drain not to the ocean, but rather to low-lying, internal basins. This is the case for the Mojave River, which feeds Soda Lake, a seasonal lake in San Bernardino County; the Owens River, which fed Owens Lake in Inyo County before it was diverted to Los Angeles; and Walker, Lee Vining, and Rush Creeks, which fed Mono Lake before being diverted to Los Angeles.

Lakes in low-lying desert basins, like Soda Lake, are referred to as alkaline because of their high salt content. They have high salt concentrations because rainfall leaches salt from rocks and soil. When streams with any amount of salt, however tiny, run into an undrained basin, the salt is left behind as the water evaporates. Several southern Central Valley rivers, including the Kern, Kaweah, and Kings, historically didn't make it to the ocean most years (and for this reason, the southern Central Valley is considered an alkaline basin). Instead, their waters were captured in the basin at the southern end of the Central Valley and created Tulare Lake, a huge tule marsh. During the wettest years the entire Central Valley would flood and connect these rivers to the ocean through the Golden Gate.

WATERSHED PROCESSES

Water that reaches a stream channel takes a gravity-driven journey that plays a significant role in shaping the physical and biological landscape. Rivers and creeks, collectively referred to as streams, vary considerably, from the smallest rivulets in the mountains to the large, broad waterways that form complex estuaries when they reach the ocean. Whether they are small tributary streams in mountain headwaters (the beginnings of streams) or reaches (any other stream segments) farther downstream in a broad river valley, streams have a number of important functions.

The particular characteristics of a stream are heavily influenced by features of its watershed. *Watershed* is a term to describe all of the land from which water drains to a common point. Watersheds are sometimes

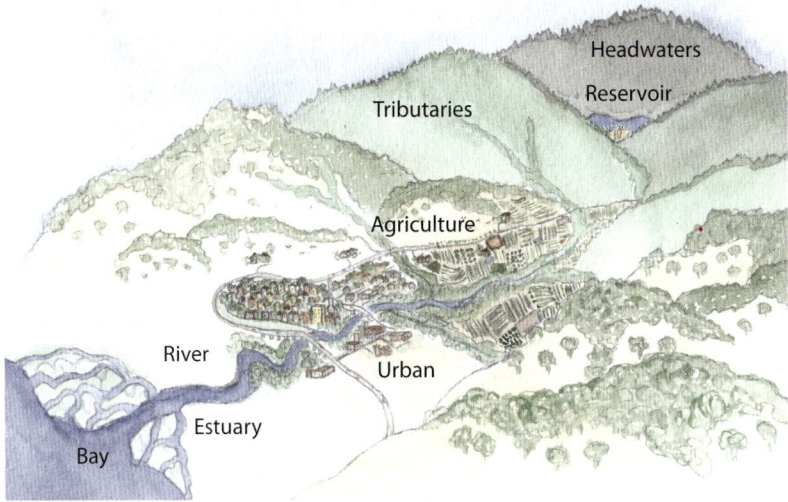

Watershed vocabulary: rain falls in the headwaters, runs from tributary streams into the mainstem river, which meets the estuary just before entering the bay and then the ocean. Watercolor by Jenny McIlvaine for the Napa County Resource Conservation District.

referred to as catchments or drainage basins, both terms referring to the idea of water being "caught" as rain or snow and draining to a common point. Think of a bathtub with a single outlet; the tub is a watershed. Your sink is a separate watershed, and when your shower water misses the tub and lands on the floor, it has moved into still a different watershed! Small streams have small watersheds, but as two small streams join to make a larger one, the stream's watershed becomes larger as well. The size of a river channel, as measured by its width and depth, is determined by the volume of water it carries, which in turn is influenced by the size of the watershed that feeds it. Essentially, the larger the watershed, the more water is gathered into the river, which creates a larger channel to accommodate the flow. Some California watersheds are huge, such as the Sacramento River watershed, which covers 27,000 square miles and carries 30 percent of the state's total surface water.

The path of a river, from its headwaters to its outlet, which is most commonly the ocean, is a complex one. Though most rivers in California run through broad valleys (such as the Sacramento, San Joaquin, Salinas, and Santa Maria), the majority of a river's drainage network, by length, is found in the headwaters. For example, more than 75 percent

Mouth of the Klamath River in June. The sandbar nearly blocking the mouth of the river changes shape intermittently. During winter high flows, it would be far less prominent. Courtesy of the US Army Corps of Engineers.

of the Russian River drainage network, in the Coast Ranges of Northern California, is composed of streams that range from 8 to 15 feet wide. These river headwaters are commonly in mountain canyons or narrow valleys, where rivers have little room to meander through alluvium. The term for streams whose pathways are limited by bedrock walls is *confined*. Confined streams are usually in the headwaters portion, and when they reach broad flatlands farther down, they become unconfined and the channel location often shifts, requiring more room to meander.

The Klamath River, however, is unusual. It is huge, draining more than 10,000 square miles, with headwaters in broad, flat volcanic plains. It then runs as a confined stream in its lower reaches as it passes through the Klamath Mountains in northwestern California.

The Klamath River notwithstanding, most rivers across the globe are oriented with confined headwaters in mountains and then unconfined reaches in valleys in their downstream portions. Geologists and geomorphologists, frequently thinking about earth-shaping processes over the long term, often refer to valleys along the lower portions of rivers as alluvial valleys, where most of the material filling the flat valleys has been transported from mountains farther upstream.

Rivers that run through alluvial valleys are unconfined by mountain walls. They can cut meandering paths through their valleys by eroding banks in places where water moves swiftly (called erosion zones) and then redepositing alluvium in slower portions of the river (deposition zones). Like landslides, erosion of river channels is a natural process, as you can see when flying over any river valley. By air, it's easy to see old sections of meandering stream channels that may not have been occupied by the river for thousands of years. Recently abandoned segments of

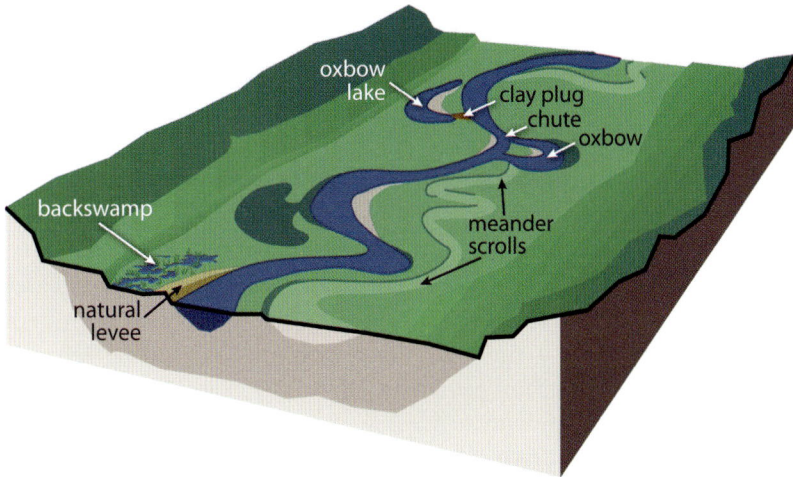

River vocabulary: rivers move in characteristic ways, creating predictable landform patterns, including oxbows and chutes, oxbow lakes, levees, meander scrolls, and backswamps. From *Stream Corridor Restoration: Principles, Processes, and Practices*, Federal Interagency Stream Restoration Working Group, 1998.

river channel may still fill with water by underground flow from the main channel and have aquatic vegetation. These abandoned channels are called oxbow lakes.

In addition to having meandering stream channels, rivers running through alluvial valleys commonly have floodplains with riparian vegetation. The riparian zone is the area adjacent to a stream or lake and is frequently characterized by specialized plants adapted to the transition from water to land. Floodplains, as the name implies, are areas that are inundated by water during high-flow events. Trees and shrubs on a river's floodplain accommodate the flooding and act as a buffer zone to surrounding drier lands. Trees such as willows, alders, and cottonwoods love having their feet in the water, and they reproduce by seed in the bare soil deposited by floods.

The inundation of floodplains has other important implications for river ecosystems. Water spilling onto a floodplain moves much more slowly than water in a river channel. As a result, very fine sediment (silt) in the water has the opportunity to settle on the floodplain. This silt contains nutrients that support plants and animals, so floodplains provide very productive habitat. The largest trees of any species usually grow in the deep, rich, well-watered soil of floodplains. For instance, the largest and tallest redwoods are found on floodplains. Floodplains also make productive farmland, and about 90 percent of the floodplain habitat in the Central Valley has been converted to agriculture. There are now attempts to restore floodplains to their original periodic flooding pattern to help recover birds like the willow flycatcher and many other California plants, birds, and insects that prefer floodplain forests.

River and stream processes provide many ecosystem services. Headwaters forests capture water and deliver it to streams and aquifers. Riparian vegetation and undeveloped floodplains absorb high water without property damage to cities and towns. Water that moves through soil and groundwater basins is naturally filtered and cleaned. Wetlands also sequester more carbon per acre than forests, making them vital for carbon storage and climate regulation. Restoring wetlands is a great example of mimicking natural processes to create solutions to human-created problems.

In fact, among the best ways to restore ecosystems is to let nature do it for us. Maintaining riparian woody vegetation and downed wood helps create channel habitat diversity including deeper pools that hold water longer through the dry season, which is important for juvenile salmon survivorship. The reintroduction of beaver (*Castor canadensis*) also improves stream habitat. Beavers were hunted nearly to extinction in America. They were viewed as pests and provided pelts that could be sold for high prices. Beaver dams are now recognized as natural water storage facilities. They slow water flow, filter sediment, augment summer flows, and establish deep pools for fish. Beaver-engineered landscapes are now recognized as natural fire breaks, and they often provide the only remaining seed bank and animal refuge after landscape-scale wildfires.

California Tribes and Native communities have long venerated the life-giving roles played by rivers, streams, and the fish within them. The return of salmon on their migratory path is central to many Indigenous traditions and ceremonies. In one notable acknowledgement of this special relationship, in 2019, the Yurok Tribal Council formally recognized the Klamath River and contributing waters and landscapes as having the same legal rights as a person.

GROUNDWATER

Groundwater is water located in air spaces beneath the land surface, in cracks and fractures in soil, rock, and sand. When it rains, some of the water percolates down from the surface and becomes groundwater. Think of dripping water onto a dry sponge. At first, the water is all absorbed. When the sponge is filled to its water-holding capacity, water begins to drip out of the sponge. In this analogy, the sponge is soil and subsurface. There are different types of bedrock, and these have different degrees of permeability, which is the capacity of material to hold water and to allow it to pass through. Some types of volcanic rock, for example, are highly porous and may hold huge amounts of water in underground aquifers. Areas of bedrock with relatively high permeability, such as sandstone, are called aquifers. Groundwater can also be fed by streams that are hydrologically connected to the aquifer. Groundwater can remain in an aquifer for days, years, or millennia, depending on its depth and location.

Aquifers are valuable because of their gigantic capacity to store water, and California relies heavily on groundwater for household use and for agriculture. In an average year, approximately 40 percent of the water used in California comes from groundwater. In a drought year, that number rises to 60 percent. According to the California Department of Water Resources (DWR), over 80 percent of Californians depend on groundwater for some portion of their water supply, and many communities are completely dependent on groundwater. In particular, many farmers rely on groundwater reserves to grow the abundance of fresh vegetables and fruit for which the state is famous and upon which the nation depends.

Groundwater is obtained from wells, which can vary in size and depth, from small domestic wells as shallow as 20 feet to enormous municipal or industrial wells thousands of feet deep. Although a well may belong to one person or farm, groundwater exists across human boundaries, and pumping from one well can therefore affect other wells. When a well is pumped, the groundwater level near that well lowers, in what is called the cone of depression, until snowmelt, rain, or a hydrologically connected stream recharges the aquifer. The rate of recharge will be dependent on many factors, such as the amount and timing of precipitation, soil type, vegetation type, geology, slope, and the size of the aquifer. If a well is pumped too heavily relative to groundwater recharge rates, the groundwater level will remain depressed. When this occurs over multiple wells within a basin for multiple years, the basin is in a condition of overdraft.

Ecosystems can be dependent on groundwater for their survival and are called, appropriately enough, groundwater-dependent ecosystems.

These areas include springs, seeps, and some types of wetlands but can also include streams and riparian areas. Groundwater-dependent ecosystems are among the most sensitive in the state.

California was the last western state to regulate groundwater, and even now, the state's groundwater storage capacity remains unknown. What we do know is that over the past decades and especially during California's extended drought from 2011 to 2017, groundwater overdraft became so extreme that some areas of the state, particularly in the Central Valley, began to experience land subsidence. The land surface actually sank! Indeed, chronic groundwater level declines in some parts of California are among the most extreme in the world.

In 2014, California started to manage groundwater for the first time, through the Sustainable Groundwater Management Act (SGMA). Ninety-four of California's 515 groundwater basins are now required to develop a Groundwater Sustainability Plan and bring groundwater use to sustainable levels by 2045. The groundwater plans are intended to prevent six conditions: chronic lowering of groundwater levels, reduction of groundwater storage, seawater intrusion, land subsidence, water quality degradation, and depletions of interconnected surface water. The plans must also address impacts to groundwater-dependent ecosystems. We can be proud that California is the first state to have specifically recognized the needs of nature in its groundwater law.

One of the most interesting aspects of SGMA is its explicit recognition of groundwater–surface water interactions. For most of the state's history, these water sources were treated as two distinct entities from a legal and regulatory point of view, though hydrologists and ecologists have long understood and documented their interaction. Groundwater flow to surface water plays an important ecological function, helping to keep creeks flowing in the dry months and providing a freshwater refuge for many species. The dynamic works in the other direction as well. During wet periods, streams can provide recharge to aquifers.

Methods to achieve groundwater sustainability range from instituting water use efficiencies, to water reuse, to desalination, to managed aquifer recharge, such as direct water injections or flooding of orchards and fallowed fields. Recharge holds high potential for increasing groundwater levels and creating long-term water supplies. Astonishingly, DWR declared in 2023, "Groundwater basins have the ability to hold 8 to 12 times more water than the state's surface water storage and reservoirs." In addition, aquifer recharge has the potential for ecosystem benefits including creating overwintering habitat for migratory birds and reducing flood risks by providing a bypass system for rivers during winter storms.

An experimental but highly promising approach is using aerial electromagnetic mapping to identify huge ancient river valleys, known as paleo valleys, under the Central Valley. These are filled with sand and gravel. Paleo valleys are ideal locations for storing diverted surface water as groundwater, as they are highly permeable and exceedingly deep and wide.

BIOLOGICAL INPUTS

A watershed provides more to its stream than just the water that falls within its boundaries. In addition to water, the watershed delivers anything else that can be transported downstream by water. These inputs may be biological, chemical, or physical, and all are relevant to stream ecology.

The insects, fish, mammals, clams, and other animals that live in a stream depend on biological inputs such as leaf litter as a food supply. These inputs can come downstream from distant uplands or from the immediate riparian zone. Biological inputs also come from the ocean to the river! When salmon, lamprey, and other anadromous fish (fish that spend most of their lives in ocean habitats but migrate to fresh water for breeding) swim upstream, they bring energy and nutrients from the ocean to the river or even deep within a forest via the scat of bears that eat them.

In addition to providing food, biological inputs from the riparian zone support other stream functions. Trees and branches that end up in a stream are called woody debris and are important to stream processes. For some organisms, woody debris provides a substrate for growth or life cycle changes. Woody debris provides a source of nutrients as it breaks down and releases carbon, nitrogen, and phosphorus to the stream. For years, land managers removed large woody debris from stream and river channels. People thought they were "cleaning" the stream. Little did they realize that streams love to be messy! This means we need to maintain and restore riparian vegetation and avoid the urge to "clean" stream channels while at the same time taking care to protect public safety and infrastructure that can be damaged as wood inevitably moves downstream during storm events. Maintaining and restoring riparian vegetation and protecting woody debris are powerful tools for restoring streams and providing cover, shelter, and feeding opportunities for the organisms that depend on them.

Invasive Species and Water Systems

Rivers and streams can act as transport systems for aquatic invasive species, allowing them to spread throughout a watershed. Hundreds of aquatic invasive species have been introduced to California—some accidentally through ballast water (quagga and zebra mussels and overbite clams), others as ornamentals (water hyacinth), for sport fishing (largemouth bass), or for commerce (nutria, hydrilla). Once here, aquatic invasives may destroy infrastructure, undermine ecosystems, and outcompete native species.

To help prevent the spread of aquatic invasives, always *Clean, Drain, and Dry* boats after each use, and never move fish or plants from one water body to another!

CHEMICAL INPUTS

As a stream moves from headwaters to lower reaches, the chemical properties of water make it well suited for transporting ionic compounds, collectively called salts. Water can weather rocks both on and below the surface; in doing so, it can dissolve compounds from rock surfaces and carry them downstream. This is the mechanism that makes desert lakes salty: streams that feed those lakes carry salts to them, and as the water evaporates each year, the weathered salts remain in the lake beds.

Other chemical inputs from the watershed have effects on stream ecology. Nutrients that are important for plant growth, such as phosphorus and nitrogen, can be weathered from soil and bedrock. These nutrient inputs are autochthonous, meaning that they come from within the watershed. They are important for the growth of plants as well as algae and other periphyton (simple organisms, sometimes referred to as biofilm, that attach to the bed of a stream). The periphyton provide the foundation for many aquatic food webs; they are the food source for some insects and other macroinvertebrates (invertebrates big enough to notice). The plants feed the primary consumers (plant eaters), which serve in turn as food for still larger organisms, including other insects, fish, and birds that turn to aquatic ecosystems for a food source.

A watershed may sometimes provide an unhealthy amount of nutrients to a stream or lake. Humans contribute to nutrient loading in streams and lakes by applying fertilizer to crops, parks, gardens, and lawns. The fertilizer is then carried to streams and lakes during storms.

Point Source and Nonpoint Source Water Pollution

Point source pollution is pollution that originates from a specific spot, such as an oil refinery. Nonpoint source pollution is what cumulatively comes from many scattered sources, such as fertilizers and insecticides from farms and yards, toxic residue from tire use, oil and grease from roads and parking lots, sediment from unpaved roads and construction sites, bacteria from animal waste and faulty septic systems, and littered cigarette butts and food wrappers from pretty much everywhere. Nonpoint source pollution is harder to control because it comes from so many different sources, so it's up to all of us to make a difference:

- Never throw refuse into storm drains.
- Pick up litter—safely, with gloves and tongs.
- Minimize or end your use of herbicides, fertilizers, and insecticides.
- Clean your car at a commercial car wash.
- Properly dispose of household hazardous wastes, including batteries, paint thinner, motor oil, and medicines. Never pour these products down any drain or toilet, and never discard them with regular household trash.
- Separate your livestock from wetlands and riparian areas.
- Clean up after your pet.
- Remember the slogan "Slow It. Spread It. Sink It." for stormwater. Advocate for your town or neighborhood to install rain gardens and filter strips.
- Collect roof runoff in a rain barrel.

Water bodies containing large amounts of nutrients are described by the word *eutrophic* (meaning "well fed," as opposed to *oligotrophic*, meaning "poorly fed") and are frequently characterized by large amounts of algae, periphyton, and aquatic plants. Eutrophic water bodies pose threats to aquatic ecosystems because periphyton and plants that accumulate in lakes eventually die and are decomposed. The bacteria responsible for their decomposition consume oxygen dissolved in the water, which can cause water bodies to become anoxic (without oxygen), meaning that oxygen levels fall below thresholds of tolerance for aquatic animals. In extreme cases, nutrient runoff can lead to dead zones. Nutrient

40% evapotranspiration

30% evapotranspiration

10% runoff

55% runoff

25% shallow infiltration

25% deep infiltration

10% shallow infiltration

5% deep infiltration

Natural Groundcover

75–100% Impervious Surface

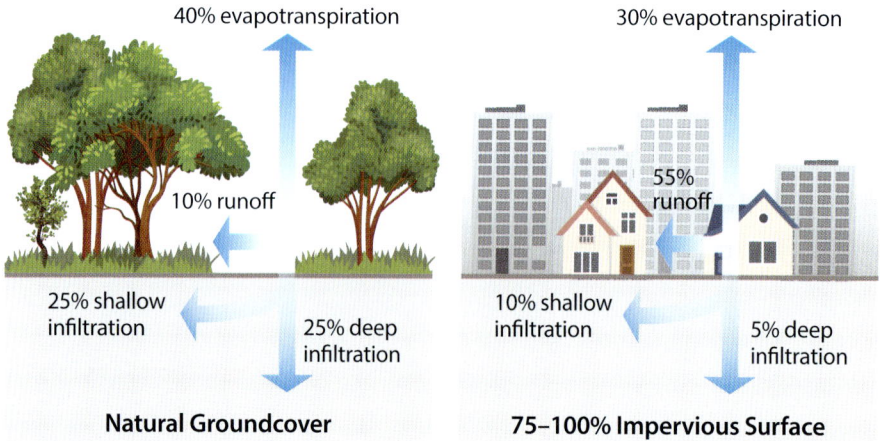

Relationship between impervious cover and surface runoff. The percentages represent an average within a wide range of results that vary with climate, amount of vegetation, ground cover, and other factors. Image from *Stream Corridor Restoration: Principles, Processes, and Practices*, courtesy of the Federal Interagency Stream Restoration Working Group, 1998.

inputs and higher water temperatures can also result in harmful algal blooms, which can kill fish and other animals and can sicken humans.

Houses, parking lots, sidewalks, office buildings, roads, and other hard surfaces pose multiple challenges to streams. These surfaces are impervious and block the natural movement of water into soil. Impervious surfaces accelerate the flow of water during storms (known as stormwater), making it more powerful and worsening flooding and erosion. Impervious surfaces create the conditions for sheet flow, which picks up pollutants such as oil, gasoline, bacteria, animal feces, fertilizers, plastics, heavy metals, and tire dust that are toxic to fish and other aquatic organisms. Sheet flow is unusual under natural conditions, but common in urban environments. Stormwater flows from hardened urban surfaces into storm drains, which carry it directly into natural water bodies untreated, thereby transporting pollutants into streams and lakes.

To address stormwater runoff and protect water quality, most new buildings in California must include vegetated filter strips, rain gardens, and other structures that slow the water down, filter it, and allow it to sink into the ground. The soil, sand, and rocks in these structures provide biological and physical filters of contaminants, cleaning the water before it reenters a stream or groundwater. This is another example of using nature for restoration, whereby the soil, rocks, and plants do the work!

PHYSICAL INPUTS

The physical material inputs to a stream play an important role in creating and changing the shape of a stream channel, and in organizing the size and distribution of stream surfaces that support many ecological processes. Geologists and geomorphologists refer to all loose minerals and rock in the river as sediment (the contrast to sediment is bedrock). *Sediment* refers to a range of size classes, from large boulders to fine silt. Sediments are delivered to streams through two mechanisms: alluvial forces and colluvial forces. Alluvial sediment (or alluvium) is material that has been moved by water, for example, during high-flow events, when channel bed materials are carried downstream by the force of flowing water. High flows may also erode the banks of the channel, taking material previously not in the stream and moving it downstream as alluvium.

Colluvium is material that reaches the stream by gravity. Most commonly, colluvium enters the stream channel through hillslope failures (also termed shallow landslides), when fine and coarse materials (soil and rocks), along with vegetation attached to them, fall into the stream. Rocky materials, such as large boulders that have been fractured and loosened from the bedrock, can enter the stream by colluvial forces but may be too large to be moved by the stream. These boulders are great for climbing or sitting on. Frequently, landslides deposit a large amount of material on one side of a stream where the landslide meets the river. This is how most rapids are formed, when a landslide deposit narrows the channel. Landslide toes may push the creek to the opposite side of the channel or fill the channel completely, causing the stream to excavate a new path.

The power of water to move objects is huge. About 90 percent of the material moved by rivers is suspended in the water. This suspended load can be composed of various materials, including fine sediment that gives high-flowing rivers a muddy appearance. The other 10 percent is bedload, the material that bounces along the streambed. Bedload includes coarse cobbles, finer gravels, and sand, as well as boulders that roll along the channel bed under very-high-flow conditions. Next time you're near a boulder-strewn river or stream, listen for cobble movement. Especially during high-flow events, you may hear cobbles and boulders bouncing along as bedload.

Though colluvial processes are considered distinct from alluvial ones, hillslope failures generally result from events in which colluvial and alluvial processes coincide. High-flow events that erode the stream channel may undercut hillslopes, destabilizing them. Here again we see the incredible power of water: its ability to undermine the stability of whole hillsides. Additionally, hillslope failures often occur when soil is saturated with water, such as during or following a heavy rainfall. Wet hillslopes become much heavier and may become unstable.

Erosion is a natural process. Problems can arise, however, when the rates of erosion increase due to land use activities and more sediment is produced than can be moved downstream by natural alluvial processes. Too much fine sediment in stream reaches can present problems for aquatic organisms. For example, in the Elkhorn Slough watershed, upland agriculture led to increased sediment transport from farms to the marsh, burying pickleweed (*Salicornia virginica*). Restoration efforts included native vegetation planting, water diversions, construction of water and sediment control basins, and stream channel stabilization, and these actions resulted in pickleweed marsh recovery. Excess fine sediments can also have impacts on fisheries, particularly for salmonids, by reducing the topography of the stream, filling in areas used for thermal and predator refuge, and raising stream temperatures.

ESTUARIES

An estuary is a mixing zone where a river meets the ocean. From a chemical perspective, estuaries tend to have lower concentrations of salt than the ocean, allowing both salt-water-dwelling and fresh-water-dwelling organisms to live in estuaries if they are capable of tolerating slightly salty water. Some organisms in estuaries, such as pickleweed, are specialized to the moderate salinity found in neither freshwater rivers nor the ocean. You can taste the salt in pickleweed if you take a bite. Nutrient inputs from land and constant mixing by tides make estuaries among the most productive of aquatic ecosystems. In an estuary you can see a mix of wildlife that came from the ocean, the river, and the terrestrial habitat. River otters and mergansers mix with seals and pelicans as osprey circle overhead.

While many lakes and rivers are eutrophic, the ocean is generally oligotrophic. In fact, the absence of nutrients in the ocean is a primary reason why the ocean appears to be blue. Alpine lakes, like Lake Tahoe, Donner Lake, and Convict Lake, are blue because they are oligotrophic. Estuaries are generally not blue. The brown and green is as much a result of biological productivity as it is of the silt and sediment the river carries to the ocean. Although oceans in general are oligotrophic, the nearshore ocean off California is enriched by upwelling. As discussed in

Gualala estuary, Mendocino County. The meandering course of the Gualala River estuary is the mixing zone for fresh water from the river meeting salt water from the ocean. Photo © 2002 P. T. Nunn.

chapter 2, upwelling happens when deep currents rise to the surface and bring up nutrients that support aquatic food chains including fish, birds, and mammals.

The dynamics of salinity in estuaries can vary according to several factors. The tides change salt concentrations throughout the day. At high tide, estuaries tend to be more saline because the tidal influx pushes salt water up the estuary. During low tide, fresh water can extend farther toward the ocean, providing the nutrients and other matter from the watershed to the estuary. The salinity of estuaries also varies seasonally. During periods of higher river flow in the California wet season, estuaries may become dominated by fresh water. The salinity of a particular location in the estuary may also vary with depth: because salt water is more dense than fresh water, it may settle below fresh water and create a vertical salinity gradient. Therefore, the species that inhabit estuaries must be able to tolerate a wide range of salinity.

Some estuaries are marked by deposition of large amounts of sediment and are referred to as deltas. Deltas are formed by stream processes, and the amount of material a stream carries depends on its speed and volume. When its speed drops to zero on entering a lake or estuary, its carrying capacity disappears and the sediments are deposited.

INTERTIDAL ZONE

As rivers reach the ocean, whether by passing through an estuary or by falling down a coastal bluff, the fresh water of the river merges with the salt water of the ocean. Along the California coast, the environment nearest the shore is called the intertidal zone. While California is 800 miles in length from Oregon to Mexico, the state has over 3,000 miles of strikingly beautiful shoreline, including beaches, piers, bluffs, rocky coves, ports, and harbors. California residents have a long history of seeking public access to beaches. The majority of Californians live within 30 miles of the coast, and billions of dollars of economic activity depends on coastal resources, including tourism, recreation, fishing, shipping, and commercial and residential development.

The intertidal zone is characterized by high energy. Anyone who has visited the seashore along California's Pacific coast will be familiar with this idea. Waves pound the shore, delivering energy continuously. This pounding is one of the defining features of the intertidal zone, and all organisms living in the intertidal zone must adapt to it, from sea stars to urchins to anemones. Two common adaptations are a hard exoskeleton (also known as a shell) and the ability to adhere to rocks.

High energy produces both positive and challenging conditions. Waves may carry sand or rocks, which erode or crush the shells of even well-protected organisms such as whelks. Waves also bring a fresh supply of highly oxygenated water, which may also be high in nutrients. The intertidal environment can be very harsh. The costs (dangers) of life in the intertidal zone are great, but these rich environments are filled with food, making it worthwhile for species to inhabit these areas and adapt to the forces of nature.

The second physical factor that characterizes the intertidal zone is the rise and fall of the tides. Low and high tides are caused primarily by the gravitational pull of the moon and, to a lesser extent, the sun, which creates bulges in the Earth's oceans as the Earth rotates, resulting in periodic changes in sea level. The effect of rising and falling water levels on organisms living in the intertidal zone is intense. They must adapt to life underwater and to life exposed to dry air and full sun! Intertidal organisms must be both aquatic and terrestrial! They may be attacked by both aquatic and terrestrial predators, from stingrays to raccoons! The other limit tidal cycles impose on intertidal organisms is that their food source may be missing for half the day. Many intertidal organisms are filter feeders. They stretch a web or net that floats in the water and traps nutrients, or, like the Pacific razor clam, they pump water through their body cavities and filter out nutrients inside their shells. In either case, when the organisms are exposed to air rather than to water, no feeding is possible.

The shoreline of California is rich in habitat diversity. Two factors that help naturalists to notice connections in seashore habitat are substrate and position (depth). *Substrate* refers to the physical composition of the shore. It may vary from fine mud in an intertidal estuary, to coarse sand on a beach, to cobbles or boulders, to bedrock on exposed headlands, and it may include combinations of these. The group of organisms that inhabits mudflats in an estuary is almost completely different from the group that lives on rocky cliffs exposed to wave pounding.

The intertidal zone is a place of fascination and wonder, with organisms that appear to our terrestrial eyes both bizarre and beautiful. If you spend time in the intertidal zone, you may see a clam spit a jet of water 3 feet in the air; have an anemone grab your finger with mobile, Velcro-like tentacles; or come across the sea clown nudibranch, a white "sea slug" with red bumps and projections, that looks ready for use in the *Star Wars* movies. The intertidal zone provides multiple ecosystem functions, including foraging and nesting habitat for birds and mating sites for elephant seals. This zone is also highly vulnerable to damage from oil spills, invasive species, and pollution, and it will be among the most affected by climate change as sea levels rise.

WETLANDS

Wetlands are transitional zones between land and water that are inundated with water periodically or permanently. Among the several types of wetlands are marshes, bogs, vernal pools, swamps, baylands, salt marshes, and riparian areas. Keep in mind that not all wetlands are freshwater. Some have high salt concentrations and support an entirely different suite of species, such as shrimp, mollusks, and pickleweed. A large number of native aquatic plants can be observed in wetlands. Pond lilies, cattails, sedges, bulrush, and arrowhead are a few examples. On a rainy day or spring evening, it's hard to ignore the loud call of Pacific chorus frogs

Vernal Pools

Vernal pools are temporary wetlands defined by a depression in the terrain with an underlying impermeable soil layer sometimes referred to as hardpan. These shallow depressions fill during the winter rains and evaporate during the summer, hence the name *vernal*, signifying spring. Vernal pools are uncommon globally because their formation requires specific soil conditions coupled with a pattern of wet winters and dry summers. California vernal pools, however, are remarkable, even within that rarified category, due to the extremely diverse array of native plants and animals these small ecosystems support, many of which are rare and endemic. Of the 200-plus species of plants associated with California's vernal pools, over 100 species are found nowhere else.

Naturalists exploring a vernal pool and associated native wetland plants in Lost Valley, Cow Mountain Ridge, Mendocino County. Photo by Kerry Heise.

coming from vernal pools and wetlands. In addition to frogs, other amphibians, reptiles, and waterfowl depend on wetlands. Prior to the conversion of most of California's wetlands, there were herds of elk, flocks of birds that darkened the sky, and grizzly bears along the shores, and mussels, oysters, salmon, and sea otters were abundant in the water.

Wetlands perform many important ecosystem functions, processes, and services, including

- Biological diversity: Wetlands provide important habitat for diverse communities of plants and animals.
- Waterfowl habitat: Wetlands are the principal habitat for migratory waterfowl.
- Fisheries: Wetlands provide spawning and rearing habitats and food supply that support both freshwater and marine fisheries.
- Flood control: Wetlands detain flood flows, reducing the size and destructiveness of floods.
- Water quality: Wetlands filter pollutants that could otherwise contaminate groundwater or the water quality of rivers, lakes, and estuaries.
- Groundwater recharge: Some wetlands recharge aquifers that provide urban and agricultural water supplies.
- Recreation: Wetlands support a multi-million-dollar fishing, hunting, and outdoor recreation industry nationwide.
- Carbon sink and storage: Wetlands absorb carbon from the atmosphere at high rates, even higher per acre than forests, and store it for long periods of time because rates of decomposition are slow.

Despite these benefits, wetlands in California have been under pressure for many years. Historically, wetlands were viewed as unattractive and perhaps useless areas and were converted by diking, draining, and filling. The San Francisco Estuary, the largest estuary on the West Coast, has lost nearly 97 percent of its historic wetlands. Statewide the number is more than 90 percent. Prior to the mid-1800s, the Central Valley had more than 4 million acres of wetland, but nearly all of it (95%) has been lost or modified.

Like the intertidal zone, estuaries and other coastal wetlands are vulnerable to rising sea levels. In addition, many wetlands are impacted by invasive species, nonpoint source pollution, and conflicts over water use and development.

There have been some positive improvements, however. In the past, rice farmers in the northern Central Valley prevented the Sacramento River from flooding their lands and then burned the stubble left at the end of the harvest. These actions had the effect of excluding huge acreages of what was once seasonal wetlands from being covered in water,

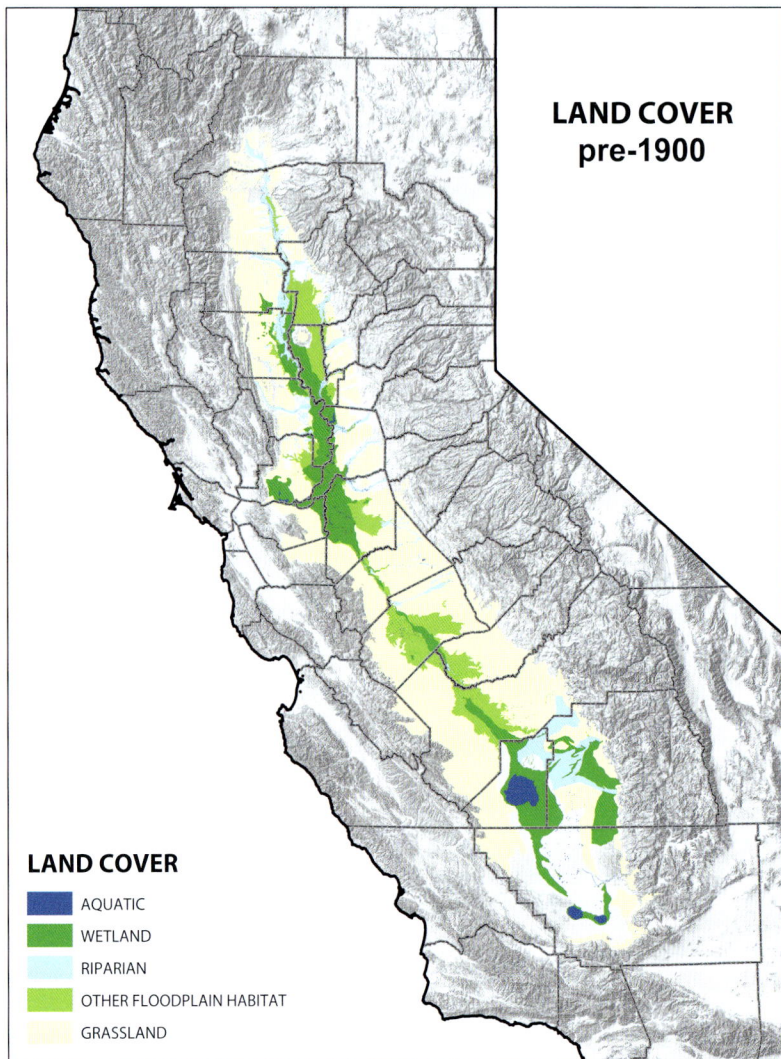

LAND COVER pre-1900

LAND COVER
- AQUATIC
- WETLAND
- RIPARIAN
- OTHER FLOODPLAIN HABITAT
- GRASSLAND

Changes in land cover and disappearance of Central Valley wetlands. More than 90 percent of Central Valley wetlands have been altered or converted for other uses since 1900. Adapted by Jason Schwenkler from maps created in 2003 by the Central Valley Historic Mapping Project, California State University, Chico, Geographic Information Center.

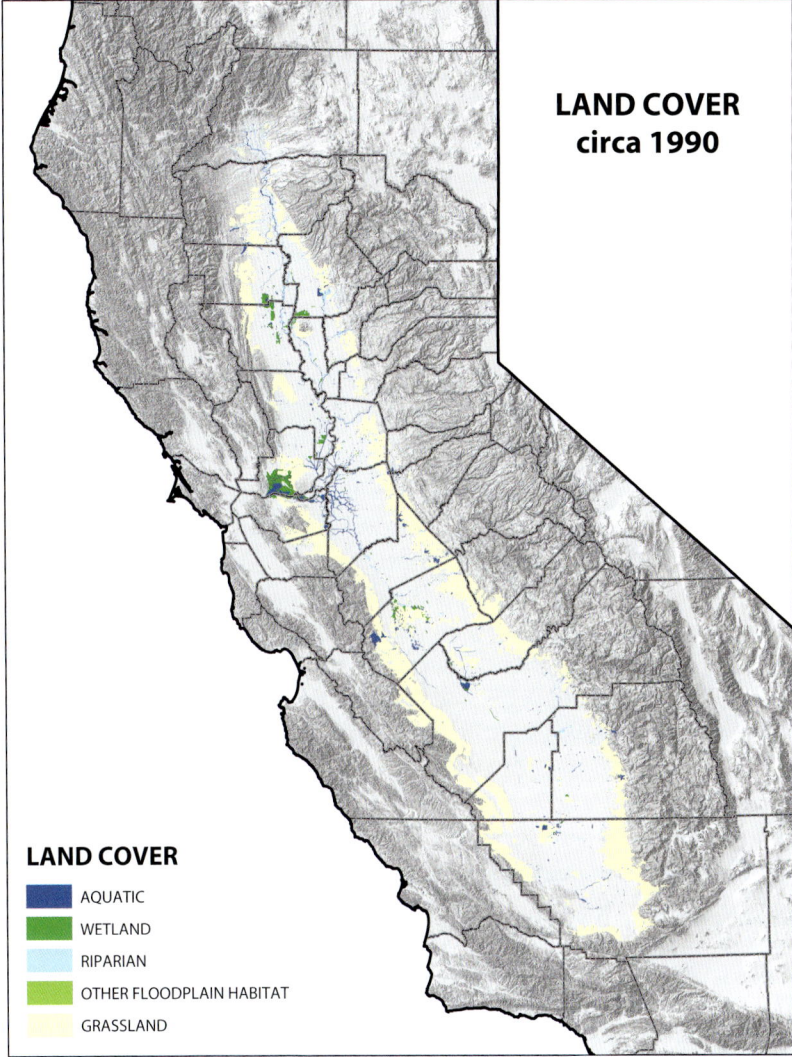

LAND COVER
circa 1990

LAND COVER

- ■ AQUATIC
- ■ WETLAND
- ■ RIPARIAN
- ■ OTHER FLOODPLAIN HABITAT
- ■ GRASSLAND

which left migratory waterfowl without winter habitat while also creating poor air quality. Pressure from state agencies caused rice growers to reimagine their practices. Now largely restricted from burning by air quality regulations, they flood their fields in winter to recycle the stubble, creating habitat for waterfowl. Ducks, geese, swans, and cranes spend the winter in the flooded fields, benefit from the leftover rice grains, fertilize the fields with their by-products, and provide a source of income to rice farmers from hunting.

CALIFORNIA'S LAKES

The topography of California generally does not lend itself to the formation of large natural lakes, but there are a few, including Clear Lake in Lake County and Lake Tahoe.

Lakes support a myriad of plant and animal species that mostly can be found in the littoral zone. The littoral zone is found at the surface of a lake or pond and receives the most sunlight. Thanks to energy from the sun, this zone can support a diverse biological community, including several species of algae (like diatoms), rooted and floating aquatic plants, grazing snails, clams, insects, crustaceans, fish, and amphibians. Dragonflies and midge eggs and larvae hang out here. Many of these species are important foods for turtles, snakes, and ducks.

California has some of the most unusual lakes in North America. Clear Lake is the largest natural lake located entirely within California and is among the oldest lakes in North America, with age estimates of 1.8 to 3.0 million years old. Approximately 10,000 years ago, Clear Lake drained to the Russian River, before a landslide blocked its outlet west and forced it to drain east via Cache Creek to the Central Valley. Clear Lake had a diverse fish fauna, with 14 native species, now diminished. One species, the Clear Lake splittail (*Pogonichthys ciscoides*), was endemic to Clear Lake but is now extinct.

Clear Lake is a naturally eutrophic lake that holds significant scientific value. Thanks to the long geological record and deep sediment layers, the lake is a treasure trove of ecological and climate information that is used to predict future environmental conditions under climate change. Clear Lake has also experienced considerable environmental degradation in recent decades, including repeated wildfires, fish die-offs, invasive species, harmful algal blooms, and mercury poisoning from open-pit mining. These issues have been exacerbated by agricultural runoff, poorly maintained septic systems, and boat recreation. Native people have lived at Clear Lake for over 12,000 years and have a strong Tribal presence today, advancing cultural preservation and Indigenous stewardship.

Lake Tahoe, nestled near the crest of the Sierra Nevada, is the tenth-deepest lake in the world, the second deepest in the United States, and among the oldest in the world. It is famed for its blueness, clarity, and alpine surroundings. Lake Tahoe is the source of the Truckee River, which flows east to a closed basin in western Nevada. Due to urbanization in the basin, channelized streams and stormwater runoff from impervious surfaces have caused Lake Tahoe to lose one-third of its remarkable transparency since 1959. Small particles of dust and sediment remain suspended in the water column for years, adding to the gradual but relentless transparency loss. Air pollution is also a factor; nitrogen pollution of the lake from atmospheric deposition is greater than from stream water input. This added nitrogen has allowed algal growth to increase by about 5 percent per year, further diminishing water clarity. Warming of Lake Tahoe due to surrounding land use as well as climate change presents a real threat to this unusual ecosystem. Both California and Nevada have embarked on efforts to restore the lake and its famed clarity.

Most naturally occurring lakes in California are much smaller than Clear Lake or Lake Tahoe. Hundreds of tiny alpine lakes are nestled in mountain cirques across the high Sierra Nevada, Cascade, and Southern California mountains. Cirques (pronounced "serks") are bowl-shaped depressions carved by glaciers. The coastal mountains of California notably have very few natural lakes.

Located in southeastern California, the Salton Sea is a unique inland saline lake occupying the Salton Sink, a depression that lies below sea level. Inland bodies of water have periodically materialized (and later evaporated) over time in the region in wetter winters. The Salton Sea became a more enduring aquatic feature when floodwaters from the Colorado River broke through an irrigation canal under construction in the Imperial Valley in 1905. Initially viewed as a recreational paradise, the Salton Sea has suffered environmental degradation due to agricultural runoff, leading to increased salinity levels, periodic fish die-offs, and high levels of pollutants such as pesticides and heavy metals. Over the same time, migratory birds have found the Salton Sea and now use the habitat it provides. Current conservation efforts aim to protect the wetland habitat and reduce exposure to toxic dust, but climate change and controversies among competing community interests make the future of the Salton Sea uncertain.

Mono Lake and Owens Lake

Mono Lake is an alkali lake in a closed basin east of the Sierra Nevada. Photo by Lisa Murphy.

Mono Lake is one of California's unique freshwater habitats. The lake evolved as a hydrologically closed basin, having no ocean outlet. Estimated to be 1 million years old, it is one of the oldest lakes in North America. It is highly saline, seven times saltier than the ocean. Water entering the lake brings salt from the mountains, and as water evaporates from the surface, the salt is left behind to concentrate. Beginning in the early twentieth century, the City of Los Angeles bought up vast tracts of land in the Owens Valley and the Mono Basin to divert water to Los Angeles for urban uses. The Los Angeles Department of Water and Power (LADWP) began exporting these waters via an aqueduct system to Southern California, an engineering marvel that led to the near-total drying up of Owens Lake and the demise of agriculture and loss of natural ecosystems in the Owens Valley.

(continued)

CALIFORNIA'S FRESHWATER FISH

California's large size, complex topography, and geographic variability have given rise to a unique assemblage of native freshwater fish species, many of which are endemic to the state. A total of 129 native freshwater

Mono Lake and Owens Lake *(continued)*

Inflows to Mono Lake were also diverted, causing lake level to drop from 6,417 feet to 6,372 feet—a decrease of 45 feet. This exposed strange underwater formations called tufa towers, and it connected Negit Island, a nesting site for 50,000 California gulls, to the mainland. The connection allowed coyotes to devastate the colony. The increased salinity resulting from these water diversions impacted the productivity of the brine shrimp and alkali fly populations, which used to emerge in tremendous numbers and were the base of a food chain supporting huge numbers of white pelicans, eared grebes, and Wilson's phalaropes.

In the 1970s, after student research demonstrated severe impacts to Mono Lake, a movement grew to reduce the diversion of water by LADWP. After many years of court battles and the establishment of the region as the Mono Basin Scenic Area, the State Water Resources Control Board issued an order in 1994 to reduce LADWP's diversions of water from the Mono Basin, enough to restore the lake to an elevation of 6,392 feet, a level last seen in 1964. This level was a compromise that accomplished several, but not all, ecosystem health objectives.

Adapted in part from Deborah Elliot-Fiske, 1995. "Mono Lake Compromise: A Model for Conflict Resolution." *California Agriculture.*

fish taxa are recognized in California. California has both anadromous fish, such as salmon and lamprey that migrate long distances between marine and freshwater environments, and resident fish species that thrive in isolated desert springs, intermittent streams, and alkaline lakes. However, in comparison to rivers of the eastern United States, rivers in California support a smaller number of fish species. This is a consequence of the harsh environment in which California's fish fauna evolved. The region has been subject to dramatic climate variation, including periods of glaciation and prolonged droughts. On an annual basis, streams in California fluctuate considerably from raging torrents in the wet season to dry beds or low-flowing trickles in the dry season. The species that have persisted demonstrate a range of local adaptations to thrive under these variable and often stressful conditions. For example, several species of pupfish persist in the few streams and pools found in California's southeastern deserts, despite extreme water temperature and salinity levels. Thus, while species richness (the total number of species) is relatively low, there is high diversity in morphology, behavior, and life history patterns in California's fish fauna.

The transformation of California's landscape throughout the nineteenth and twentieth centuries due to mining, farming, and population growth has led to dramatic declines in freshwater fish populations. Over 80 percent of California's native freshwater fish species are vulnerable to extinction over the next 100 years. The decline in native fish fauna has been caused by a wide range of human activities, including dams, water diversions, habitat modification, and introduction of nonnative species. Dams disrupt connectivity within river networks, preventing the migration of salmon and isolating fish populations above dams. Dams and diversions also alter the natural flow regime of rivers, reducing or eliminating peak flows, artificially enhancing summer flows, and changing cold-water rivers to rivers with warm-water conditions.

The loss and modification of freshwater habitat is another challenge to California fish. The alterations to river corridors by land use conversion, flood protection infrastructure, and riparian vegetation removal have altered habitats and natural ecosystem processes on which native fish depend. The Sacramento–San Joaquin Delta was once an enormous tule marsh with meandering channels providing excellent habitat for native fish. It has been largely drained for farmlands protected by levees along dredged river channels. The Los Angeles River now flows for 50 miles within a concrete-lined channel. Virtually all forested watersheds in Northern California have been altered by logging and hydraulic mining. Hydraulic mining permanently degrades rivers by obstructing their ability to meander.

At least 50 nonnative freshwater fish species have been introduced in California and have had widespread negative effects on the state's native fish through hybridization, predation, competition, and disease. Most introductions of nonnative fish were deliberate attempts to provide sportfishing opportunities or for aquaculture, although accidental introductions of aquarium fish have occurred. The introduction of fish and invertebrates from ballast water in cargo ships is a growing problem, and some refer to the San Francisco Bay as the "most invaded estuary in the world." The success of introduced aquatic species in California is associated with modifications of aquatic habitats: nonnative species thrive in artificial reservoirs, drainage canals, and regulated rivers. A dramatic example is the Russian River. In the early 1900s, an interbasin hydroelectric diversion began sending water from the Eel River to the Russian River, changing the natural flow regime of the Russian from a flashy, cold-water system with disconnected pools by the end of the summer to a perennial, warm-water river with abundant introduced species. The Russian's once-renowned winter steelhead run has been reduced to a small fraction of its former glory.

Based on current trends, much of California's native fish fauna is likely to disappear in the next century. The need for freshwater ecosystem restoration and conservation is critical. The multifaceted and complex causes of fish population declines require that freshwater conservation measures be integrated with water management and land use planning.

RIVERS TODAY

California rivers have been altered substantially in the past 150 years. Natural water features, such as marshes, and natural processes, such as floods, are seen as problems for human development. In an effort to prevent flooding, provide storage, and move water from where it rains (the north) to where agriculture and population centers are located (Central and Southern California), people have built a complex system of canals, dams, reservoirs, pumping plants, levees, tunnels, and aqueducts. We have drained huge areas of wetlands, and irrigated land that is normally dry or desert.

California has been called the most hydrologically altered landmass on the planet. Whether or not this statement is correct, California undoubtedly has an unusually complex and highly engineered system of water conveyance, the most notable examples of which are the Central Valley Project, the California State Water Project, and the Colorado River Aqueduct.

In recognition of this crazy quilt of engineered water conveyance and to highlight potential vulnerabilities in the systems, these projects and the many smaller ones that compose California's water world are sometimes called the water grid (similar in concept to the power grid).

Once a vast tidal marsh, the San Francisco Bay/Sacramento–San Joaquin Delta (the Bay-Delta) is now a key center of water supply linkages, transportation infrastructure, and agricultural productivity in California. Where the Sacramento and San Joaquin Rivers meet, the Bay-Delta begins, establishing the eastern boundary of the San Francisco Estuary, the West Coast's largest estuary.

The Bay-Delta provides drinking water for 25 million Californians, and it delivers irrigation water subsidized by both California and national taxpayers to a $27 billion agricultural industry. The Bay-Delta harbors 57

Major Water Projects in California

The Central Valley Project (CVP) was begun in 1938 and is operated by the US Bureau of Reclamation. The CVP is composed of 20 reservoirs, 11 power plants, and approximately 500 miles of canals and aqueducts, carrying 7 million acre-feet of water from as far north as Lake Shasta (north of Redding) to the Central Valley and portions of the Bay Area, as well as generating 5.6 billion kilowatt-hours of electricity annually.

The State Water Project (SWP) was built in the 1960s and is operated by the Department of Water Resources. The SWP includes 22 dams and over 700 miles of canals, tunnels, and pipelines. The SWP transports water (via the California Aqueduct) from the Bay-Delta to Central and Southern California and provides water to 27 million people and 750,000 acres of farmland. The SWP includes the world's tallest water lift: at the Tehachapi Mountains, 14 pumps carry water 1,926 feet up and over the Transverse Ranges.

The Colorado River Aqueduct is part of a larger system distributing water from the Colorado River to seven states, numerous Native American Tribes, and Mexico. California's allocation is 4.4 million acre-feet a year for both agricultural and urban uses, the latter administered by the Metropolitan Water District of Southern California.

major reclaimed islands, 1,100 miles of levees, and hundreds of thousands of acres of marshes, mudflats, and farmland, making the entire system hydraulically, ecologically, economically, and socially complex. Land subsidence, sea level rise, earthquake risk, development, and ecosystem degradation are looming threats to the benefits the Bay-Delta provides.

More than 700 plant and animal species, some unique to this system, are supported by the Bay-Delta. For bird-watchers, the estuary provides ideal viewing of waterfowl as many migratory birds feed and rest there as they pass along the Pacific Flyway. The Bay-Delta once supported some of the state's largest fisheries, and many people are working to bring back winter- and spring-run Chinook salmon (*Oncorhynchus tshawytscha*), Central Valley steelhead (*Oncorhynchus mykiss irideus*), delta smelt (*Hypomesus transpacificus*), Sacramento splittail (*Pogonichthys macrolepidotus*), and southern green sturgeon (*Acipenser medirostris*). Invasive species now pose a threat to native populations in the Bay-Delta, however, as the estuary has experienced the arrival of more than 250 alien aquatic and plant species.

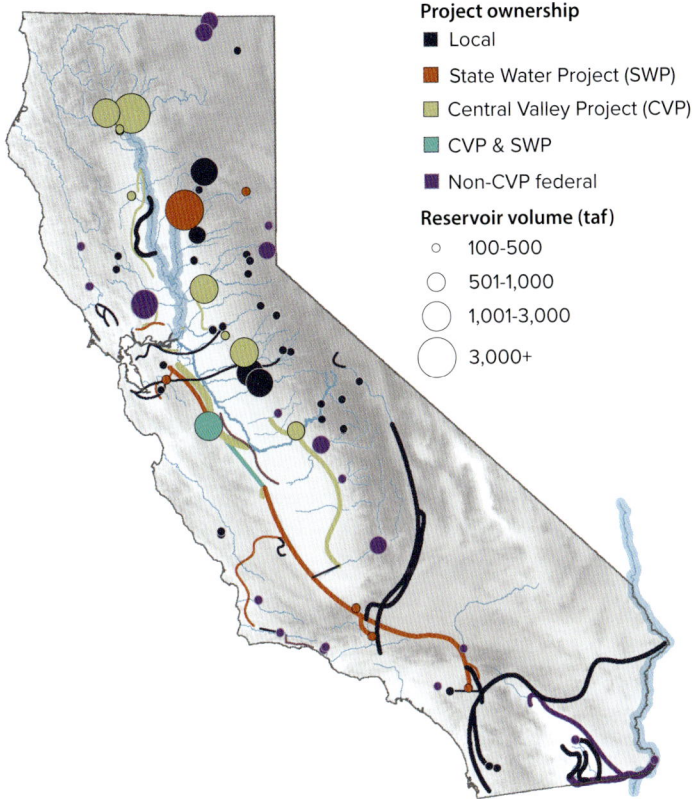

California's major surface water storage and conveyance systems. The lines show rivers in blue and show built conveyance infrastructure facilities (canals, aqueducts) colored by ownership. Circles are surface reservoirs with storage capacity greater than 100 taf (thousands of acre-feet), scaled to size. California's water grid includes both surface water and groundwater, though groundwater basins are not shown here. Figure excerpted from *California's Water Grid* by Alvar Escriva-Bou, Ellen Hanak, and Jeffrey Mount, Public Policy Institute of California, 2019.

Management of the Bay-Delta in the twentieth century addressed water supply, flood control, irrigation, power production, and navigation. Today's efforts to restore the Bay-Delta require attention to improving water supply reliability and to protecting, restoring, and enhancing the Bay-Delta ecosystem, which necessitates large multiagency coordinated planning efforts. Proposed plans have various names, ranging from the peripheral canal to the Delta Conveyance Project, which would ship water around or under the Delta from the Sacramento River to the southern part of the state. All the plans have been controversial. Changes in salinity levels in the Bay-Delta resulting from new management regimes could negatively affect biota and agricultural uses. Still, the debate highlights the need for long-term planning and policy setting for the region and for the state.

To prevent flood damage, levees have been built throughout the state, usually earthen walls along riverbanks to prevent rivers from spilling onto floodplains. Levees disrupt the natural process of flooding and the deposition of sediment and nutrients on the floodplain, and they threaten the survival of organisms that depend on periodic inundation. Levees are not a permanent solution, and over time, levees degrade and break, especially during high-stream-flow events.

Dams are another piece of the water grid puzzle. Dams can provide protection from flooding, store water for drinking and irrigation, and be a low-carbon source of electricity. However, dams damage rivers ecologically by completely altering their natural flow regimes. Changing a stream's natural flow regime can change its shape, the type and size of sediment accumulating in the riverbed, the type of fish and other aquatic life that can be supported, and the type of plants found along the river. In addition, dam building in California historically resulted in forced displacement of Indigenous Peoples and the submersion of Native American villages, burial sites, and sacred sites—sites that are often biodiversity hotspots.

When dams promise protection from flooding, floodplains are encroached upon for agricultural and urban uses, and the dams are relied upon to preserve human infrastructure. For example, Folsom Dam regulates flow in the American River and protects the city of Sacramento from flooding. While providing flood protection, these dams alter the magnitude of high-flow events. Water stored behind dams can also be released at times when higher water would not naturally occur, affecting the timing of flow. Typically, when dams are managed for agricultural purposes, water is stored during the wet season, then released to farmers in the dry season.

Some large dams also have diversion channels operated by irrigation districts to distribute water to farmers. For example, the Friant-Kern and Madera Canals divert water from the San Joaquin River stored behind Friant Dam. Similar diversions are operated for municipal drinking water; Lake Mendocino in Mendocino County and the Hetch Hetchy reservoir in Yosemite National Park are both drinking water sources.

Like levees, dams are not permanent structures. All dams fill with sediment and diminish in storage capacity. In addition, aging dam

Shasta Dam on the Sacramento River north of Redding. Courtesy of the US Bureau of Reclamation.

infrastructure has a tendency to eventually fail, as the Oroville Dam spillway did in 2017.

As the environmental costs and the risks associated with dams become more evident, the era of dam construction is passing and the era of dam removal is arising. Highly successful fish restorations due to dam removals in California include those in Butte Creek in 1993 and the Carmel River in 2015.

The largest dam removal in US history took place on the Klamath River in 2024. Four dams were initially constructed for electricity generation and to control the timing of water release. These dams blocked spawning runs of Chinook and coho salmon, steelhead, and Pacific lamprey. The removal of the four dams was the result of years of collaboration among Tribes, environmental groups, fishing advocates, Pacific Power, and state and federal agencies. The Hoopa, Karuk, Yurok, Shasta, Klamath, and Modoc Tribes were the primary advocates for the dam removal, fighting for many years for restoration of the fisheries their lifeways have long been dependent on. They were joined by many environmental groups advocating for river health and fish restoration.

Removal of the dams opened 400 miles of historic stream habitat for spawning lamprey, steelhead, and salmon. Despite that, the removal does not completely restore the system, as dams and diversions remain in the headwaters area, threatening the shortnose sucker and the Lost River sucker. Results from other large dam removals, such as on the White Salmon River (2011) and the Elwha River in Washington (2012–2014), however, show that river recovery can be rapid and dramatic.

In addition to the dozens of large dams in California, thousands of small reservoirs speckle the landscape. Small reservoirs are commonly designed to detain water for one property, often for cattle. Many such reservoirs are located on small headwater streams. In California, these streams generally flow intermittently during the wet season and flow very little, if at all, during the dry season. Because they are designed for individual properties, they are generally composed of earthen fill and do not have capacity to release water until they are filled, at which time they overflow into their stream channels. Because it stores a small amount of water relative to the overall discharge of a larger drainage network and is on a small stream, a small reservoir on a headwater stream may not impact anadromous fish, but the cumulative impact of many small reservoirs can have substantial effects on the whole river. Together, they may withhold enough water to significantly reduce the discharge in a creek that salmon or steelhead need for spawning and rearing. The many small reservoirs speckled throughout the Coast Ranges and foothills also alter habitat by providing water sources in an otherwise summer-dry environment. Invasive species like feral pigs and bullfrogs are aided greatly by these water sources.

The impact of decades of alterations to the hydrological landscape has left California's land less able to absorb and retain the rain that falls and to ensure that native species have adequate water for survival, leaving everyone vulnerable to the impacts of drought and in some cases flood. Riparian, wetland, and bay restoration projects provide tools for reducing problems and improving conditions for the future. Movements such as the 30×30 initiative create targets for California, the nation, and the world to conserve 30 percent of lands and coastal waters by 2030 and move toward 50 percent by 2050. For water, it will be important to conserve headwater forests, restore wetlands and riparian zones, advance stormwater capture and bioretention basins, improve forms of flood control such as horizontal levees, and move infrastructure out of floodplains when possible.

Restoration of the Los Angeles River is a hopeful example. For decades, flood control agencies attempted to "control" the river by channelizing it in concrete. More recently, efforts to restore the river are becoming a reality and include riparian habitat and wetland restoration, daylighting of tributaries, removal of invasive vegetation, and allowing the river to reoccupy its floodplain. These actions have the multiple benefits of maintaining flood management through natural processes that also encourage wildlife, improve recreation, and increase infiltration.

CALIFORNIA WATER USE

The management of water resources and water use is one of the greatest challenges facing the people of California today: we must meet the demands of a huge population and a sizable agricultural economy while maintaining the sustainability of aquatic ecosystems and equitable access to water.

The State divides water use into three primary categories: environmental, agricultural, and urban. If all forms of agriculture (irrigated cropland, pasture, orchards) are lumped together, agriculture represents the single largest water-consuming industry in California, accounting for 40 percent of all fresh water used in an average year. Environmental uses account for 50 percent, and urban water use for 10 percent. (Statistics that cite agriculture as using 80% of water refer to 80% of developed water.) Environmental use, the majority of which occurs in Northern California, refers to water that is left in streams and wetlands for ecosystem benefits and to protect endangered species.

Water restrictions are a perennial issue, and California communities are continually looking to augment freshwater supplies through a wide range of approaches. Some are as old as civilization and include rainwater harvest and water conservation; others are more technically complicated, such as desalination, treated wastewater, or stormwater reuse. Some are a combination of the two; for example, municipal water suppliers are increasingly providing apps to customers to help them track their water use and receive alerts for suspected leaks.

CALIFORNIA WATER LAW AND WATER EQUITY

It may seem strange to dedicate so much space to exploring human-constructed water systems and law in a natural history book, but you cannot understand water in California without recognizing that artificially constructed systems and their impacts are foundational to the civilization we have created here. Much of the drinking water consumed in Los Angeles is supplied from the Feather River in Northern California. Almost none of the water consumed in Los Angeles comes from the

LA basin. Similarly, almost all the water drunk in San Francisco comes in a pipe from Yosemite National Park in the Sierra Nevada mountains. In California, where water is located and what it is used for are usually not the products of nature. Whether any given person's well is dry, whether water is available for fish passage, how resilient the state's remaining wetlands are, and how well they will be able to absorb rising sea levels are often determined by these constructed systems.

California water law is infamous for being complicated, yet it underlies issues concerning water security and who has access to water. Since statehood in 1850, California has controlled the right to take water from streams for legally defined beneficial uses. The right to divert water from surface water bodies is overseen by a complex system of rules that combines two basic principles: the riparian doctrine (riparian rights) and the doctrine of prior appropriation (appropriative rights).

Riparian rights allow for "reasonable and beneficial use" of water immediately adjacent to (that is, physically touching) a landowner's property. The water must be used right away; it cannot be stored for more than 30 days, but the user does not lose the right from lack of use. Appropriative rights, on the other hand, allow the rights holder to take water from one place and use it somewhere else, as well as store it for later use. This right can be lost if not used.

Appropriative rights can be divided into senior rights and junior rights. Senior water rights are defined as those that were embedded in law prior to 1914. Those with senior water rights have virtually untouchable and unlimited access to their water. Senior water rights holders can fully exercise them for defined beneficial uses before water is available to any junior water rights holders. Junior water rights come in various varieties, depending on when they were inscribed in law, and can be more easily curtailed by the state in times of drought.

The doctrine of prior appropriation has as its basic principle "first in time, first in right." Essentially, whoever declared their right to the water first got it in perpetuity. In the 1850s, the declaration process was literally to nail a notice to a post or tree. The principle of first in time was designed to benefit settlers to the detriment of California Tribes, many of which had been forcibly removed from the land beforehand.

The implications of California's foundational water rights policies are profound and still felt today, separating Tribes not only from land and waters considered central to cultural traditions, but also from the ability to secure riparian or senior rights to water for community needs. Native Americans were first in time, and some Tribes hold or have secured their rightful highest-priority water rights. However, many Tribal communities in California do not have access to water rights on their current landholdings or within their homelands. Hence many are working through the courts to secure water rights and improve their access to fresh water.

The criteria used to determine whether a water right can be granted have changed over the past few decades. Many of these changes came about because the definition of *beneficial uses* now includes ecosystem functions and state-defined Tribal Beneficial Uses, and these uses are considered equal to other uses, such as for agriculture and cities. For example, water needed to provide habitat for fish is now considered a beneficial use. This means that in some cases, water rights to a stream can be exercised only if enough water is left in the stream to provide conditions necessary for fish to survive. Some of these cases are further complicated by federal laws to protect endangered species, which provide additional protections of specific in-stream flow conditions. In the 1980s, the California Supreme Court also decided that the State Water Resources Control Board had to consider the public interest in preserving natural resources when granting water rights permits. The legal recognition of public trust resources was the foundation for restoring water to Mono Lake.

Another challenge with the water rights system is that many streams are overallocated, meaning that more water has been "promised" to rights holders than exists in the system. Although it is outside the California water rights system, an important and egregious example of overallocation is the Colorado River Compact of 1922. Seven states have rights to Colorado River water, to the tune of 15 million acre-feet (maf) of water a year (plus another 1.5 maf for Mexico), yet according to the Water Resources Research Center at the University of Arizona, "data from three centuries indicate an average flow of about 13.5 maf. Also, flows are highly erratic, ranging from 4.4 maf to over 22 maf." California has a water right for 4.4 maf, although historically it has diverted more than that.

Changes to California water rights are afoot, primarily driven by recent droughts that have made it clear that California is hamstrung by its water rights system, including the ability of the state to curtail water use and the ability of rights holders to share available water with others.

In addition, an equally important effort is being made to reexamine the entire water rights system through an equity and justice lens. In parallel to the larger LandBack movement, these efforts seek to recognize the theft of Tribal land to gain access to surface water, and the flooding of Tribal communities to create the reservoirs that support the California water grid.

WATER CHALLENGES FOR THE PRESENT AND FUTURE

It is clear that climate change will alter the water future of California. We can expect extended periods of drought alternating with years of too much precipitation. This greater climate variability has implications for aquatic biota. Warmer temperatures will increase the percentage of precipitation that occurs as rain rather than as snow. More rainfall and less snow will shift the timing of flows, from snowmelt yielding water from April through September, to rain events discharging from November through March. Even if the same amount of annual discharge from the mountains occurs, a shift from snowmelt to rain is likely to reduce the amount of water available for urban and agricultural use and increase the frequency of damaging floods.

Policies and proposals for meeting our future water needs have been coming at record speed. New laws have mandated water conservation as a way of life. New dams have been discussed. The Public Policy Institute of California sums up the situation succinctly: "The [water] grid is ill-prepared to handle climate change. . . . Yet the grid is also the state's most valuable asset for adapting to the changes in store."

The water supply for the future will include a suite of options both regional and local: water recycling, agricultural and urban water use efficiency and conservation, drip irrigation, improved soil management for water-holding capacity, stormwater reuse, gray water, and desalination. The Pacific Institute estimates that aggressively pursuing conservation and reuse strategies could yield the equivalent of 10.8 million to 13.7 million acre-feet of additional water per year.

Water is a life-giving and life-sustaining compound that has been brilliantly and devastatingly manipulated in modern California. As water and ice shaped the California landscape, water management, transport, and ownership shape our social organization. With extreme variations across the state in rainfall and water availability, California has been a leader in many areas of water management and conservation. In a future where soils will be drier and rainfall less predictable, our ability to adapt will be crucial. Our state is known for public engagement and innovation, two assets that will be important in the search for a sustainable future. One thing that Californians can count on is that water will remain at the center of economic, ecological, social, and policy discussions far into the future.

Explore!

VISIT A PERMANENT OR TEMPORARY WETLAND

Visit a wetland or a vernal pool and notice the changes in the plant communities near and far from the water. Document changes that you think might have occurred since European settlement. Look at an aerial photograph and observe in the field to see if you can determine whether the size, shape, or extent of the wetland has changed over time. For example, have some areas been filled for development? You can find wetlands by accessing the US Fish and Wildlife Service National Wetlands Inventory website.

LEARN MORE ABOUT WATER ISSUES

Water management and law in California are the subjects of volumes of books and articles, and the information presented here is greatly synopsized. Naturalists are encouraged to read more, such as *Cadillac Desert* by Marc Reisner, to get a more complete picture of the history, economics, and politics of California water. There are many fantastic blogs, such as Maven's Notebook or the California Water Blog, with up-to-date California water news in an easy-to-read form.

JOIN A CREEK OR BEACH CLEANUP

Creek and beach cleanups are fun and meaningful. Find organizations near you that hold these events. Start with the annual Coastal Cleanup Day. You can find out more on the California Coastal Commission's website.

PARTICIPATE IN COMMUNITY WATER QUALITY MONITORING

This is so much fun, you won't believe it yields important results. Volunteer for a Clean Water Team, and learn to identify and sample benthic macroinvertebrates or measure dissolved oxygen, pH, turbidity, and other water quality measures. The State Water Resources Control Board has a list of teams throughout the state at their Clean Water Team website.

EXPLORE WHERE YOUR DRINKING WATER COMES FROM

You can learn a lot about California water by finding out where your water comes from. If you get water from a small or municipal water system, find out about its sources of water and the water rights that enable access to those sources. If you own a well, find out if you are in a basin with very low, low, medium, or high priority, and for the last two, learn about your local Groundwater Sustainability Plan. Compare your results with those of a friend who lives 10 miles away. You may be surprised by the differences.

4

Plants

Plants surround us, shelter us, feed us, and form the aesthetic and functional backdrop to our lives. Were it not for green plants, there would be no life as we know it on Earth. Despite their central importance to our lives, many of us are no more connected to the plants we live among than we are to billboards. Ask yourself these simple botanical questions: How many kinds of trees can you name on sight? What is the first bush to bloom in your neighborhood each spring? Are the conifer trees on Figueroa Mountain near Santa Barbara the same species as those you can find in the Sierra Nevada? Are they the same species as the trees in the mountains above Los Angeles? Plants can be categorized into various groups or sets that share characteristics, life style traits, and adaptations to their environment. This chapter starts with huge generalities about all plants and gradually focuses in on specific connections to some of the important plants of California.

LIFESTYLES OF RICH AND FAMOUS PLANTS

Warm and wet conditions promote the growth of plants. Cool, cold, and dry conditions cause plants to slow down metabolically. Thus, tropical rainforests are "ideal" spots for plants to grow. Most of the world, however, is not a tropical rainforest, so the story of plant adaptation is the story of the development of alternative lifestyles to cope with

Vernal pools are hotbeds of plant speciation, and many vernal pool plants are specialized for only that habitat. This vernal pool landscape near Proberta (Tehama County) has abundant Mima mounds in the ground separating the pools. Mima mounds are small, domed hills in very old, stable landscapes. They may be the result of pocket gopher burrowing or wind accumulation. Photo by Evan Keeler-Wolf, www.ekwolfphotography.com.

the diverse deviations from ideal climate conditions that the majority of the world presents.

Some plants live for a few months, others for many years. Indeed, the longest-lived organisms on Earth are plants. Aspen trees send up new shoots connected to mother trees by roots, and a huge stand can develop from a single seed! Some of these large clones are estimated to be thousands of years old. Trees are perhaps the most familiar, archetypal plants. Trees generally grow from seeds, starting out as collections of a few cells and enlarging to become the most massive and the tallest organisms on Earth. Trees are said to be both perennial and woody. Perennial plants are those that live more than one year and generally pass an unfavorable season in a dormant or inactive state.

Trees can also be divided into those that are deciduous and those that are evergreen. Deciduous trees and shrubs lose their leaves during the unfavorable season, whereas evergreen trees and shrubs may keep individual leaves for more than one growing season, and may be photosynthetically active during the unfavorable period.

Shrubs are another type of woody plant. They generally have more than one stem, whereas trees often have only one main trunk. Shrubs generally keep to below 15 feet tall; trees are often much larger. Is the 25-foot-tall, single-trunked rhododendron in your neighbor's yard a tree or a shrub? Exceptions to these rules abound, so *shrub* and *tree* are general terms without precise distinctions.

The last category of woody plants of relevance to California naturalists is lianas, or woody vines. Most plants desire to maximize their exposure to sunlight. Essentially, they compete for sunlight. Trees compete by putting huge allocations of resources into growing a stem that allows them to overtop other plants to be in the sun. Lianas are cheaters: they utilize the structure of trees and shrubs to get to the canopy without making the investment in a large stem. There are four genera of lianas that are common in California: poison oak (*Toxicodendron*), wild grape (*Vitis*), honeysuckle (*Lonicera*), and virgin's bower (*Clematis*). These four plant genera illustrate the three mechanisms lianas use to climb: twining, tendrils, and adventitious roots. Honeysuckle and virgin's bower stems wrap around the stem they climb; they twine. Poison oak can climb tree trunks or bare walls using sucker-like roots: adventitious roots. Grapes use the grappling hooks of the plant world: tendrils.

Annual plants are distinct from perennials in that they complete their life cycle in one growing season. They pass the unfavorable season as seeds. Those of you who have grown tomatoes from seed will be familiar with the concept. Tomatoes have an annual lifestyle: make the most of one growing season and leave behind seeds. Such well-known spring wildflowers as farewell-to-spring (*Clarkia*), baby blue eyes (*Nemophila*), and popcorn flower (*Plagiobothrys*) are annuals, as are many of the plants that grow wild in the hills of California.

The California poppy is a familiar wildflower. Poppies are herbaceous (nonwoody) plants, and they have a little secret: they grow as both annuals and herbaceous perennials! In favorable circumstances they switch from being an annual to being an herbaceous perennial. At the close of the favorable season, poppies may move carbon compounds and minerals to their roots. At the return of the favorable season, they may begin to grow once again, powered by the energy and nutrients stored in the roots, rather than starting from a seed. This phoenixlike rebirth driven by stored reserves is a major class of adaptation of California plants, referred to as the herbaceous perennial lifestyle. Examples of other plants with this lifestyle that may be familiar to you are soap root (*Chlorogalum*), blue dicks (*Dichelostemma*), and mule ears (*Wyethia*). Trees and shrubs are easy to tell from herbaceous plants. Annuals and perennials are more difficult to distinguish. Often you will need to expose the root or follow the plant's growth over the course of a year to distinguish an annual from a perennial.

The stored energy in a mariposa lily (*Calochortus*) bulb or a snakeroot (*Sanicula*) tuber allows it to produce leaves much more quickly than the seed of an annual poppy, which must start from scratch to build a plant body by photosynthesis. Thus, herbaceous perennials can overtop and exclude annuals in stable, predictable environments, like woodlands and shrublands. But in open and disturbed environments, annuals come into their own, capitalizing on their ability to produce numerous seeds. Human activity, in maintaining many open and disturbed environments, has created a bonanza for annuals.

PARTS OF A PLANT

Stems and Leaves

Plants grow by making stems. In order for a plant to produce a leaf, it must make a stem. Some stems are very compact (think of a cactus), others more elongate and easily visible, but all plants must make stems to produce leaves. Leaves are usually the primary locality of photosynthesis, thus it behooves plants to make leaves. As the stem elongates, leaves are produced sequentially, from the base of the stem out to the tip. At the base of each leaf is a remarkable structure, a bud. A bud is a clump of tissue specialized to be ready to grow a new stem, leaf, or flower. Buds are often visible at the base of the leaf. The easiest time to see buds on deciduous plants is during the dormant season. Often buds in this season are enlarged and prominent, but even during the growing season and on evergreen plants, buds are often discernible. Buds consist of the tissue that will become the new stem, leaf, or flower, surrounded and protected by one or many bracts or scales. The bracts and scales are evolutionarily modified leaves. They enclose the dormant bud tissue and protect it from weather extremes (cold, dryness) and predators. When the new stem begins to grow, the bud scales relax and open to allow the new stem to elongate. Typically, they fall from the plant and may be noticeable on the ground, or the bud scale scars they leave upon dropping may be visible or prominent on the stem.

As stems elongate, they produce successive leaves along their length. Leaves come in a startling variety of shapes and sizes and are arranged upon the stem in typical patterns. Leaves may be arrayed along a stem in alternate, opposite, or whorled fashion. The leaves themselves may be simple (undivided) or compound (divided). Leaves can also be very thin—pine needles are in fact a kind of leaf!

Leaf forms: examples from five California native plants. From left to right: red elderberry (*Sambucus racemosa*), an odd-pinnately compound leaf with toothed leaflets; California bay (*Umbellularia californica*), simple leaves with entire margins, leaves alternate on the stem; snowberry (*Symphoricarpos albus*), leaves simple with entire margins, leaves opposite; oceanspray (*Holodiscus discolor*), leaves simple with toothed margins, leaves alternate; columbine (*Aquilegia* sp), a three-parted compound leaf with shallowly lobed leaflets, leaves largely arising from the base of the plant. Photo by Kerry Heise.

Flowers

Flowers can be thought of as variations on a theme. The theme is four whorls of modified leaves: the sepals, petals, stamens, and pistil. Very generally, the function of the sepals is to enclose and protect the inner parts: the petals, stamens, and pistil. Petals are the often highly colored part of the flower that signals to potential pollinators that "there may be a reward here; please investigate" and can be thought of as a kind of advertising. The stamens are the functionally "male" parts; they are the site of pollen production. The pistil is the functionally "female" part, largely characterized by an ovary within which the seeds develop.

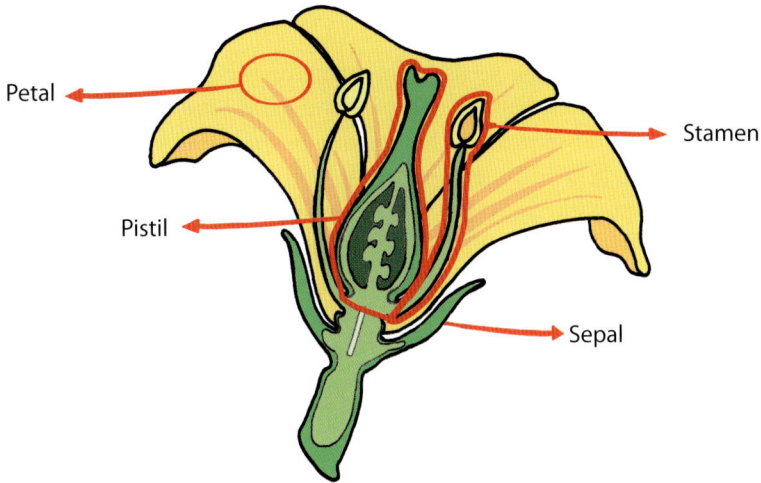

Parts of a flower. The critical parts every naturalist must be able to recognize are the sepals, petals, stamens, and pistil. Courtesy of Mariana Ruiz Villarreal.

The function of flowers is to produce seeds. The function of seeds is to allow the plant to reproduce, both locally and distantly. The intermediate step between flowers and seeds is the fruit. Fruits are the enlarged, mature ovaries that were present in the flower. Fruits come in a bewildering array of shapes and sizes, from tiny orchid fruits an inch long with 25,000 seeds to the baseball-sized fruits of the buckeye (*Aesculus*) with only one seed. The unifying theme is that all fruits are ripened ovaries.

Because plants have had millions of years to try different strategies for success (about 300 million years in the case of flowering plants), many different solutions to the mechanics of seed production have evolved. Every type of flower you see has a unique, ancient lineage. The result is a bewildering array of different-looking flowers (about 300,000 kinds on Earth today). In some plants the sepals and petals are just alike and are called tepals (many lilies). Different parts of the plant may be modified to produce nectar: the leaves (*Prunus*), petals (*Ranunculus*, *Aquilegia*), sepals (*Ipomoea*), stamens (*Viola*), or the base of the pistil.

What this means is that every flower in your garden or your local park is a mystery waiting for you to investigate. Try dissecting a flower. Perhaps you'll start with a California poppy. Look for the sepals, petals, stamens, and pistil. Try to find an old poppy flower that has dropped its sepals, petals, and stamens, to see what the ovary looks like as the seeds develop.

Pollination

Functionally, producing seeds is the name of the game. To reproduce, plants need to put staminate and pistillate parts together. That is, they must unite pollen with ovaries. For maximum effectiveness (that is, for the best rearrangement of genetic material), it is best to mate with another individual, not with yourself. (Yes! Plants have the option to mate with themselves!) Most plants attempt not to mate with themselves; they try to promote outcrossing. One of the best methods plants have developed to ensure outcrossing is to use animals to carry pollen from one plant to another. This system is a typical business deal, with payments (rewards) for services rendered, advertisement, and competition. Think of a bumblebee hovering around a manzanita (*Arctostaphylos*) flower in January. Why should it? For the reward. Manzanitas pay bumblebees with sugar water (nectar). Some flowers pay in nectar, others pay in pollen, a very few pay in oils, but there is no free lunch. To some extent everyone gets what they need or want. The bumblebee (pollinator) gets food for itself or its offspring. The plant gets transfer of pollen from one flower to another.

Plants that use animals to transfer pollen must perform two functions: advertising and reward. They must attract the pollinator and reward it for its work. The advertising generally consists of gaudily colored sepals or petals or stamens. This is a visual attractant to cue the pollinator. Olfactory cues (smells) may also be used, either separately or in combination with visual cues. This is one of the great ancillary benefits of pollination: wonderful smells are produced that we can enjoy if we take time to smell the roses. There are, however, plants that are pollinated in a different way.

Some plants are pollinated entirely or in part by wind or water transport of pollen. Grasses are perhaps the most familiar examples of wind-pollinated plants. Because wind-pollinated plants do not need the services of animal transport agents, they can dispense with the costs of advertisement (pollinator attraction) and reward (pollinator payment). A quick look at a grass will show that most do not have colorful parts or smell good to attract an animal. A closer look would show that most grasses have dispensed with sepals and petals completely, and they do not have parts that produce nectar as a reward.

Seed Dispersal

One of the most fascinating aspects of plant form and function is the complex array of solutions that have developed to transport seeds to

How to Identify a Plant

If you are walking your child to school and see a really cool tree, how can you find out its name? That will depend on why you want to identify the tree. If you are trying to get to know the plant and want to develop an intimate, personal relationship with the tree, you might use a dichotomous key to figure out its name. The advantage of this approach is you are sequentially asked to choose between two options. Each couplet challenges you to look more closely at the plant, to examine it in ways you might have overlooked had you not been asked. This is a slow, careful method that will provide satisfying results and learning over the long run. If, on the other hand, you simply want to be able to comment to friends about how amazing a buckeye fruit is, or you want to be able to look it up and read about it, there are quicker ways. You can get applications for your phone that will use a photo you snap to give you suggestions for what plant you are looking at. UC California Naturalists use iNaturalist, a smartphone application that can help you identify a species, allows experts to review your observation and help verify it, and once it is verified, provides researchers with information to expand our understanding of the species distribution and habitat. Another easy-to-use phone app for identifying plants is Seek. Uploading observations can help you learn and connect with other naturalists. Once you have a name, you might want to go to the Calflora website and look up this plant to see a map of the distribution of the species in California, which is really handy for knowing if the plant naturally occurs in your area.

suitable places for them to grow. Seeds can be transported externally when they are caught on the fur or feathers of mammals or birds. They can also be transported internally, as when bears eat manzanita berries and the seeds are pooped out in a great steaming heap on the other side of the mountain, or when birds eat elderberries (*Sambucus*) and the seeds fall from the sky or a perch to splat on a landslide. One of the most wonderfully whimsical seed dispersal systems you will see is dispersal by ants. Many plants in California, and more globally, produce seeds with an oil-rich food body (an elaiosome) attached. Ants gather the seeds, eat the elaiosome, and discard the seed, which then grows in its new home. California examples of ant-dispersed plants are *Dicentra, Scoliopus, Trillium, Vancouveria*, and *Viola sempervirens*.

PLANT COMMUNITIES OF CALIFORNIA

Every individual plant is unique in its form, and each species has a distinct set of environmental tolerances that influence where it can be found. When plants with similar environmental requirements are regularly found together, we call them a plant community. Such communities blend into one another, so the concept is useful mostly as a human construct to make it easier to talk about natural environments.

The plant communities of California are determined by the various plant species' genetic requirements and tolerances interacting with all of the aspects of the local environment. Climate, geology, and interaction with animals—pollinators, dispersers, and predators, notably elk, cattle, and people—influence the local and regional distribution of plants. Fire is also a huge factor in determining the distribution of plants. The broad treatment of plant communities that follows generalizes the plant communities of California into only a few, easily identifiable groups.

Beach Vegetation

Beach vegetation, or strand, is the low, sparse, windswept carpet of plants occupying the sandy shore of the Pacific Ocean, and the dunes and bluffs immediately adjacent to them. This group of plants is composed primarily of annual or herbaceous perennial plants. These plants must be able to put up with the thrashing delivered by ocean waves, high salinity, extreme winds, shifting substrate, and blowing sand. Many of these plants have very wide geographic ranges, with seeds that are dispersed by flotation in ocean currents up and down the Pacific coast. This community is highly altered by people in many places, with seawalls, coastal homes and communities, and introduction of European beach grass being the main avenues of impact.

Grassland

There is an abundance of grassland in California, mostly composed of nonnative Eurasian annual grass species, which often exists without much shrub or tree cover. The occurrence of grassland throughout low-elevation California is intimately intertwined with historical land use, although there are some naturally treeless areas. Woodlands, forests, and chaparral have been cleared throughout the state for agriculture or pasture, usually by cutting and burning. The natural succession of grassland to brush to woodland or forest is retarded or suspended by grazing or burning. In most areas, grasslands return to shrublands and woodlands in the absence of fire or grazing. Wind-dispersed coyote brush (*Baccharis pilularis*) and bird-dispersed poison oak (*Toxicodendron diversilobum*) are usually the first colonizers, followed by trees such as oaks (*Quercus*), bay trees (*Umbellularia californica*), or Douglas fir (*Pseudotsuga menziesii*).

This phenomenon is illustrated by the Golden Gate National Recreation Area. The land between the Golden Gate Bridge and Olema was a series of cattle ranches between 1820 and 1970. The ranches were first logged, then heavily grazed. Photos of the coastal slope from the 1950s and 1960s show expansive grasslands (cattle pastures). Since the land was purchased by the National Park Service in the 1970s, it has not been grazed. Coastal scrub has almost completely covered the former pastures and is now being invaded by Douglas fir trees and bay trees, in part due to fire suppression.

A few areas where native grasses thrive still exist. Rangeland managers and restoration ecologists would like to replace exotic annual grasses with native grasses, as this increases the native biodiversity associated with these areas. In nonnative annual grasslands, removing livestock and fire allows rampant production of thatch and results in the exclusion of native grasses and forbs (small broadleaf plants).

Many otherwise rare grassland species are found primarily on serpentine soils. These are generally unfriendly places for plants to grow unless, like some California native plants, species evolve to thrive in these conditions and thereby avoid competition with species less tolerant of these harsh soils. Serpentine soil areas support a higher ratio of native plants to introduced plants than do nonnative grasslands. Areas with serpentine soils are great places to see incredible displays of brilliantly colored wildflowers.

Salt Marsh

Salt marsh is another community of herbaceous plants, those adapted to periodic inundation (flooding) by salt water, followed by exposure to extreme solar radiation. Salt marsh is further characterized by extremely low soil oxygen and high rates of soil deposition, thus the plants of salt marshes must be able to cope with regular inundation and burial. Salt marsh plants are twice daily inundated with salt water and must be able to cope with high soil salinity, as well as low oxygen. Not so obviously, many salt marshes have creek channels running through them and may seasonally be flooded with fresh water, another extreme rigor to which marsh plants must be adapted. When you think of salt marshes, you want to include both the vegetated and the bare areas (the mudflats, tidal channels, and stream channels). The bare mudflats and the vegetated areas are intimately linked and function together as parts of a system, with the vegetated areas occupying slightly higher elevations and performing primary production, and the bare areas at slightly lower levels, getting inundated more deeply, and supporting diverse communities of invertebrate consumers and decomposers.

Despite all of these seeming disadvantages, salt marshes are sites of very high primary plant productivity and behave as nurseries for fish and other inhabitants of the estuary environment. Salt marshes are typically very flat and occupy areas of high utility for human beings. Thus, salt marshes suffer greatly during development. Ninety percent of the original Bay-Delta salt marshes are now converted to other uses. Los Angeles and San Diego have similar histories of salt marsh loss. Salt marshes are perhaps the most threatened plant community in California.

Freshwater Marsh

Freshwater marsh is an assemblage of different microhabitats. It includes the giant tule beds of the primeval Central Valley, the complex sloughs and islands of the Delta, vernal pools imbedded in grasslands, the wooded transitional marshes at the upstream edges of salt marshes, and the woody and herbaceous vegetation surrounding lakes. Anywhere seasonal or permanent standing fresh water is a primary factor in determining what plants grow on a site can be considered a freshwater marsh.

The plants of freshwater marshes, like those of salt marshes, often must cope with low soil oxygen. They also experience seasonally fluctuating water levels. Seasonally or permanently flooded areas exclude the great majority of plants and provide an opportunity to those that can cope with these conditions.

Prior to the damming, diking, and channelization of the Sacramento and San Joaquin Rivers and most of their major tributaries, these great streams flooded much of the Central Valley annually during the wet season. Controlling these rivers has made possible one of the greatest, most productive agricultural enterprises in history, and it has eliminated huge, expansive, productive wetlands. When the big rivers of the Great Valley flooded each year, they filled many off-channel basins with water to depths of 3 to 12 feet for 3 to 6 months. As the floods receded, the

The Life of the Salt Marsh Bird's Beak

Flower stalk of salt marsh bird's beak (*Chloropyron maritimum*). The leaves are purple and covered with white crystals of excreted salt. Photo courtesy of Brad Kelley.

One of the unique inhabitants of salt marshes of the Bay-Delta is the salt marsh bird's beak (*Chloropyron maritimum*), an inconspicuous annual plant known only from salt marsh habitats in coastal California. The salt marsh bird's beak has seeds that are stimulated to germinate when salinity is low. In high-rainfall years, when the "salt" marsh is mostly rinsed of salt by flows of fresh water, the bird's beak sprouts in abundance. In years of low rainfall, when salinity remains high, the bird's beak may sprout in very low numbers or not at all! The plant completes its life cycle (9 months) in "typical" salt marsh conditions as the rains pass, tides bring salinity levels back up, and day length increases.

basins gradually dried down, so by the end of September they were baked hardpans of dry, cracked soil. These basins were populated with humongous stands of tules (*Schoenoplectus*).

Another unique feature of the large, flat river valleys of California is the development of small, shallow basins (vernal pools) on a variety of terraces and low spots in the valleys and surrounding foothills. These basins fill with water a few inches to a foot in depth, dry gradually, and support a unique flora of annual plants growing in concentric rings determined by water depth.

Anywhere a river or creek enters a lake, a salt marsh, or a larger river, fresh water may pond seasonally or permanently. These "transitional marshes" are often wooded with willows (*Salix*), alders (*Alnus*), ash (*Fraxinus*), maple (*Acer*), or sycamore (*Platanus*). Beneath these deciduous trees you will find the herbaceous vegetation of a marsh: cattails (*Typha*), tules, sedges (*Carex*), and rushes (*Juncus*).

All of these marsh habitats are extremely important to wildlife, from the great flocks of migratory waterfowl found in seasonal wetlands of the Central Valley; to beavers, muskrats, and rails in transitional marshes; to the endemic delta green ground beetle (*Elaphrus viridis*) that occurs only in association with Central Valley vernal pools in Solano County.

Shrub Communities

Coastal scrub and chaparral are shrub communities, characterized by dense stands of shrubs 3 to 15 feet tall. The shrubs are usually close together and intricately branched enough to make human passage difficult or impossible. Bears, deer, pigs, coyotes, gray foxes, wood rats, and rabbits, however, move through them with ease. The two main types of brush communities in cismontane California are coastal scrub, near the ocean, and chaparral in hotter, drier, more interior sites. In the interior (chaparral) the herbaceous layer is often depauperate, although near the coast (coastal scrub) the herbaceous plants between the shrubs may form the majority of the biomass.

Chaparral is often composed of nearly pure stands of manzanita (*Arctostaphylos*) 8 to 15 feet tall. In other areas, the cover is a mix of manzanita, chamise (*Adenostoma fasciculatum*), wild lilac (*Ceanothus*), scrub oak (*Quercus*), and bush monkeyflower (*Diplacus aurantiacus*). Chamise and bush monkeyflower commonly form pure stands. Chaparral occurs on hot, south-facing slopes and on hillsides characterized by impoverished soil such as heavy clay or thin, rocky soil. Chaparral is often the aggregation of woody plants that will first colonize a disturbed area, especially after a fire. Manzanita seeds are known for their ability to remain viable through long periods of dormancy (50–75 years!). Mature stands of chaparral provide a shaded seedbed for their successors, oak woodland and coniferous forest. Chaparral shrubs, especially chamise, provide excellent deer browse, and their growth is often retarded by the "hedging" effect of this browsing. Coastal scrub generally lacks manzanita and chamise, rarely rises above 8 feet in height, and is usually dominated by coyote brush, coastal sage (*Artemisia californica*), California blackberry (*Rubus ursinus*), flowering currant (*Ribes sanguineum*), and bush monkeyflower.

Mixed Evergreen Forest

Mixed evergreen forest is a composite of many different trees, including oaks, bay, Douglas fir, madrone (*Arbutus menziesii*), tan oak (*Notholithocarpus densiflorus*), and others. This forest forms dense stands with adjacent trees touching each other's canopies. Mixed evergreen forests occupy sites subject to frequent or infrequent, often intense, fires. Ability to stump sprout after fire, as well as having leathery, hard, often spiny, evergreen leaves, are the prime attributes of this community type. All plant communities intermix with other community types to a greater or lesser extent, but botanists do recognize a wide variety of community types and species associations. The mixed evergreen forest grades into and borders conifer forest, oak woodland, and grassland. In many places distinguishing them is somewhat arbitrary. Old forests and ancient trees are difficult to find in this community type, since the stands are very fire-prone, and individual trees often succumb to the fires that exclude conifers from the site.

There is a very odd word you must learn if you want to be a California naturalist: *sclerophyllous*. The word comes from the Greek roots *sclero* ("hard") and *phyllon* ("leaf"). Sclerophyllous plants are those with hard leaves, often with short internodes and marginal spines. The hard, evergreen leaves of coast live oak (*Quercus agrifolia*) are classically sclerophyllous, as are those of the other evergreen oaks, madrones, and bay trees and many of the shrubs that occur in chaparral. The needles of pine trees and Douglas firs can also be considered sclerophyllous. Sclerophyllous plants are thought to have developed in response to increasing aridity. Over time (20 million years), as California became increasingly dry, many species of plants were excluded—that is, they went extinct or were extirpated from California. Others changed; they evolved in response to the changing climate and developed adaptations to resist drought. Hard, evergreen (sclerophyllous) leaves is one such adaptation. The waxy coating on many sclerophyllous leaves, reduced numbers of stomates, and having stomates only on the lower surface of the leaf all help minimize water loss through the leaf surface. These abilities are critical in the Mediterranean climate California shares with Chile, the Mediterranean

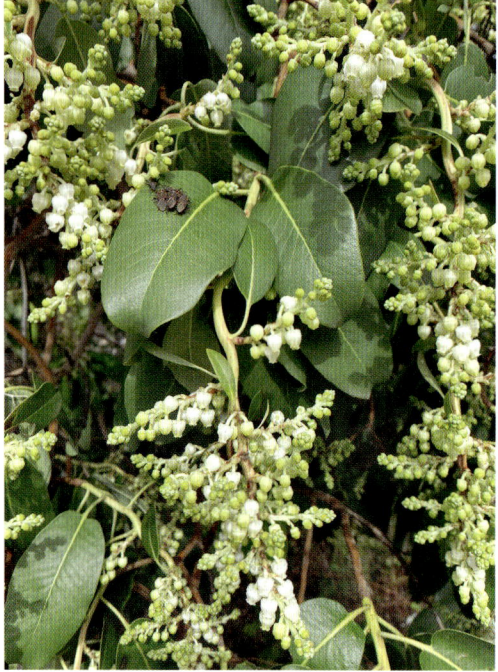

Box elder bugs on the sclerophyllous leaves of a blooming madrone tree (*Arbutus menziesii*). Sclerophyllous leaves are hard, waxy, and usually simple. Photo by Greg de Nevers.

region, South Africa, and southwestern Australia. All of these other areas with Mediterranean climates also have vegetation types dominated by sclerophyllous plants, so if you learn the concept here, it will serve you well in your travels to these other Mediterranean climate areas.

Oak Woodland

Oak woodland is perhaps the most widespread and characteristically "Californian" plant community in California. The term is usually reserved for the lands covered in deciduous oaks, while hillsides clothed in evergreen oaks fall under the banner of mixed evergreen forest. Oak woodland is rich and variable in its species composition, shares dominance among five tree species, and has the richest, most diverse shrubby and herbaceous understory of any forest type in California. Oak woodland also seems to harbor the greatest variety of native perennial herbs of any community.

California's oak woodlands are amazingly rich biologically. They provide habitat for 120 species of mammals, 147 species of birds, 60 species of amphibians and reptiles, and 5,000 species of insects and arachnids. Astonishingly, 1 acre of oak woodland can contain between 10 and 100 million individual insects and other invertebrates! In addition, oak woodlands provide watershed protection, open space, and opportunities for recreation.

Oak woodland can be dense or savanna-like. On cool north-facing slopes, the deciduous Oregon oak (*Quercus garryana*) is often abundant in Northern California. It is the only California oak that continues north to Oregon, Washington, and British Columbia, where it seems to appreciate the increased rainfall. Oregon oaks can be thought of as our most moisture-loving oak. Blue oak (*Q. douglasii*) is the dominant oak on many extremely hot, arid sites and in fact forms a "bathtub ring" around the Central Valley, essentially mapping the rain shadow areas in three dimensions. It is the diametric opposite of Oregon oak, specializing in the most arid sites that support large trees. Valley oak (*Q. lobata*) is abundant on deep soils of flats at low to medium elevations. It grows to enormous proportions in the deep, rich soils of valley bottoms, especially along creeks. Valley oaks also have the ability to survive flooding, often growing in areas where water stands for months during the wet season.

Oak woodland provides a variety of habitats for other plants, including those with deep shade, light shade, sunny openings between widely spaced trees, and seasonal shade below deciduous trees, as well as branches where epiphytic lichens and mosses perch. Perhaps the seasonal change of light penetration into these deciduous forests is the most salient characteristic, accounting for the preponderance of herbaceous perennials and shrubs in the forest understory.

Once established, oaks become highly resistant to fires. Thick bark allows California oaks to withstand even moderately intense fires with little damage. When fire destroys foliage or small branches, oaks are capable of epicormic sprouting: they force out new shoots from larger branches and re-create their small branches and leaves quickly. When the entire trunk and major branches are killed, oak trees will often sprout from the burned stump and roots. They are highly fire-resistant trees, occurring in a landscape that has evolved with frequent fires as one of the signature conditions of life.

Riparian Forest

Riparian forest is found along riverbanks and other bodies of water. The floodplains of the Sacramento, Salinas, and Los Angeles Rivers and other large and small rivers and streams are often lined with deciduous trees. The names of these trees read like a catalog of eastern deciduous forest genera: *Acer, Alnus, Fraxinus, Platanus*. These are genera that do not do well in the hot, dry hills of California. Essentially, these trees are hiding in the cracks, escaping the severe drought that descends on California every June by living on the banks of streams with their toes in the water. None of these riparian trees have the adaptations to aridity so widespread in other California plants. They don't need to! They have their roots down in a permanent water source that allows them to transpire as much as they want to, with more water immediately available to replace that which was lost. Riparian forests provide flood control, habitat, stream bank protection, and erosion control.

Coniferous Forest

The coniferous forest is characterized by evergreen cone-bearing trees (conifers) with one or more species sharing dominance. In much of California, the Douglas fir is an important component of different coniferous forest communities. In places, especially in the fog zone along the coast, Douglas firs share dominance with the unique California relict (survivor) called the coast redwood or sequoia (*Sequoia sempervirens*). At the drier extremes, Douglas firs cohabit with drought- and fire-adapted conifers like the knobcone pine (*Pinus attenuata*) and the Coulter pine (*P. coulteri*). They may also interdigitate (interlock) with or border the coastal pine called the Bishop pine (*P. muricata*). Douglas fir trees live in dynamic equilibrium with the plant communities that surround them. Douglas firs constantly throw seeds into mixed evergreen forests, oak woodlands, chaparral, and grassland in an attempt to take over these communities' land holdings. Douglas fir trees will take advantage of the partial shade provided by oaks or manzanitas, seed beneath them, and grow up through them. The Douglas firs then shade out the other species, eventually killing them. Fire interrupts and/or reverses these changes, as young Douglas firs are killed outright by fire while oaks sprout from trunk or stump and manzanitas sprout readily from seeds or burls. Large (60-plus years in age), thick-barked Douglas fir trees are resistant to ground fires, but they succumb to crown fires.

Fire Pines

There is a class of coniferous trees in California that illustrates how closely adapted to fire the flora of California is. These trees, called closed-cone pines or fire pines, produce cones (woody, seed-bearing structures) that don't open the first year. Knobcone pine is one example of the closed-cone pines. The cones and the seeds they contain are stored on the tree, often for many years.

When a fire enters a stand of trees, the closed-cone pines burst into flame, actually encouraging the fire with flammable chemicals stored in their needles and bark. The entire stand is burned, the trees are killed, but the fire has another effect. It allows the cones to open, releasing the stored seeds, which sprout in the now-open and fertilized habitat following the fire, and the stand is renewed. What this leads to is even-aged stands of trees over large patches of landscape. The danger of this lifestyle is that if fire returns too frequently, before the fire pines mature and make cones, the stand may burn a second time and be lost. An opposite adaptation is exhibited by the Santa Lucia fir (*Abies bracteata*), a California endemic that may be the rarest fir in the world. The Santa Lucia fir, with its thin bark, avoids fire by growing on steep, rocky north-facing slopes where fires rarely start or spread.

Redwoods

Redwoods are one of the signature trees of California. Virtually everyone has heard of them. They are a major attraction, bringing tourists from around the world. Most Californians have seen and perhaps walked in a redwood forest, and many of us have them growing as cultivated ornamentals within a block or two of our homes. Redwoods are conifers, like the Douglas fir and the fire pines. Redwoods are also a paleoendemic species—a species with a range that is a small remnant of a former much larger geographic distribution. Redwoods have a long evolutionary history and a great fossil record. Fossils of redwoods have been found far beyond the current, relictual range of redwoods. Redwood fossils are known from throughout the western United States and Canada, as well as from Europe, Greenland, Alaska, and China. The oldest known redwood fossils date back 160 million years!

Desert Shrublands

California's hot deserts, the Mojave, Colorado, and Sonoran Deserts, located in the southeast portion of the state, are primarily shrub-dominated landscapes. The three deserts are distinguished by their climates. The Mojave Desert, located farthest north, is influenced

Adaptation to Diverse Conditions

The Sierra Nevada range provides a clear example of how elevation and other environmental factors, such as latitude, aspect, parent material, and soil, determine the types of trees that grow in a certain place. On the west slope of the Sierra, forests generally arrange themselves in broad elevational belts dominated by one conifer or several tree species. Imagine a hike from the Central Valley floor, somewhere east of Fresno, to the top of the Sierran crest. You would start in oak woodlands and grasslands of the lower foothills, which give way to conifers such as Douglas fir (*Pseudotsuga menziesii*), white fir (*Abies concolor*), ponderosa pine (*Pinus ponderosa*), and sugar pine (*P. lambertiana*). As you climbed higher, these species would be replaced by forests of red fir (*A. magnifica*), Jeffrey pine (*P. jeffreyi*), and lodgepole pine (*P. contorta* ssp *murrayana*). Higher still, you would find yourself in a subalpine forest of mountain hemlock (*Tsuga mertensiana*), western white pine (*P. monticola*), whitebark pine (*P. albicaulis*), foxtail pine (*P. balfouriana*), and limber pine (*P. flexilis*). The subalpine forests consist of short trees shaped and pruned by the heavy snow and high winds of their high-elevation habitat.

primarily by winter rain from the Pacific. The hotter, subtropical Colorado and Sonoran Deserts receive moisture from convection storms during the summer months (July–September), in addition to winter storms from the Pacific Ocean. Creosote bush (*Larrea tridentata*), white burr sage (*Ambrosia dumosa*), and brittlebush (*Encelia farinosa*) form vast areas of scrub vegetation across all these deserts, especially in well-drained washes, bajadas, and alluvial fans.

One of the characteristics people love about deserts is the presence of succulent plants. These are plants like cacti, with compact stems and few or no leaves, and of diverse evolutionary backgrounds, that survive in arid areas by storing water in specialized tissues. A distinctive feature of the Mojave Desert is woodlands of a small tree with succulent leaves and succulent stems, the Joshua tree (*Yucca brevifolia*). One of the charismatic shrubs of the Colorado and Sonoran Deserts is the semisucculent ocotillo (*Fouquieria splendens*), whose spiny stems are reminiscent of cacti and whose tubular red flowers attract hummingbirds.

Diverse woodland in the mountains of the eastern Mojave Desert. The flower stalks belong to the succulent *Nolina parryi*. Photo by Greg de Nevers.

Mosses, Liverworts, and Hornworts (Bryophytes)

Mosses, liverworts, and hornworts compose a group of land plants called bryophytes. Mosses and liverworts are the plants that form the green carpets often seen on rocks and on the trunks and branches of trees. Bryophytes are frequently epiphytes (they grow on other plants), though they also can be epipetric (growing on rocks), can live on soil, or can even be aquatic, as are some hornworts. Bryophytes evolved a growth strategy where a vascular system is not needed! Imagine your body with no blood vessels. Bryophytes simply absorb moisture directly through their leaves. Bryophytes are, therefore, small plants, rarely more than 8 inches long. Most moss and liverwort leaves are only one cell thick, and they rarely reach 0.4 inches in length.

Bryophytes reproduce not by seeds, but by spores. The reproductive part of the bryophyte produces a capsule with thousands of spores. Many bryophytes can also form new plants by vegetative means.

Bryophytes have features that make them rather unique among land plants. Since bryophytes lack roots, they remain attached to the substrate by simple, hairlike projections called rhizoids. Bryophytes help

The terrestrial fire moss (*Funaria hygrometrica*) comes up in abundance following wildfires. Photo by Greg de Nevers.

protect soil from erosion, and they gradually release water and nutrients stored in their cells back into the environment. They can deal with long periods of dryness by shutting down all activity and then coming back to life rapidly when water again becomes available. They possess the ability to live in two very different worlds—dripping wet one day, dry the next. They survive well on rocks and in deserts, though they are only active when moisture is present. Mosses only live in places that are moist, at least seasonally, and so they are sometimes called "the amphibians of the plant world."

Bryophytes occur in every habitat, from bare rocks to deserts to mountaintops. They are especially common in temperate and tropical rainforests. They can form massive colonies that carpet the landscape. Peat mosses (the genus *Sphagnum*) play an important ecological role in cold temperate regions, especially in the Northern Hemisphere, forming massive peat bogs that sequester large amounts of carbon. Dried peat is used for heating, like firewood, and is important in the nursery trade to increase planting soil's water-holding capacity. Historically, sphagnum moss was used to dress wounds, due to its

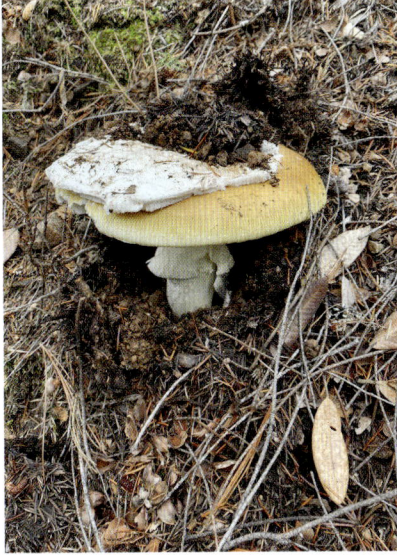

The coccora (*Amanita calyptrata*) pushes up through duff. Notice the white universal veil remnant on top of the cap, and the partial veil on the stalk. Photo by Greg de Nevers.

antiseptic properties. Peat moss gives scotch whisky its distinctive smoky flavor. In California there are 20 species of *Sphagnum*, each with only a few scattered occurrences. If you want a real puzzle, look at the distribution maps for *Sphagnum* species on Calflora (www.calflora. org), and ask yourself how this pattern could arise! Bryophytes share a very ancient common ancestor with flowering plants, and they may be a good stand-in for imagining what the first plants to colonize land might have looked like.

Many groups of plants have members whose common names include *moss*, but reindeer mosses are actually lichens; Spanish moss is a pineapple relative, a flowering plant; and sea moss and Irish moss are algae, so be careful not to associate these with real mosses.

Fungi

Fungi are strange organisms. They are not plants; they do not photosynthesize. They are heterotrophs; they depend on other organisms to gather food the fungi can consume. They eat an amazing variety of substances: Saprophytic fungi consume only dead things like leaves or elk dung. Parasitic fungi attack living things like trees, flies, or wheat. Some fungi, like the oyster mushrooms (*Pleurotus*) commonly seen as white shelves on alder trunks, are active predators, snaring nematodes (tiny invertebrates), anesthetizing and consuming them. Fungi consist of elongate, tubular, stringlike cells called hyphae. It is these hyphae that form the white filamentous mats (mycelium) often attached to mushrooms people dig up in soil or manure. The white netlike mycelium is the actual body of the organism, while the "mushroom" is the reproductive part (the "fruiting body"). Unlike plants, fungi do not make flowers; they reproduce with spores, as mosses do.

Mycorrhizal Relationships

One of the little-appreciated roles fungi play, and perhaps the most important ecologically, is as symbiotic partners with plants. Mycorrhizal fungi, that is, the fungi that form these partnerships with plants, penetrate the roots of trees or wrap their hyphae around them. The tree and the fungus then exchange materials. Generally, the tree gives the fungus the products of photosynthesis (carbohydrates, which are food for the fungus). The fungus aids the tree's uptake of water and nutrients, including nitrogen, phosphorus, zinc, and copper. Mushroom hyphae are far finer in diameter and much greater in number than tree roots, so the hyphae have a much larger surface area. This allows for a more intimate, penetrating relationship with soil particles than tree roots have, and thus greater contact with water and nutrients than tree roots can establish alone. This type of symbiosis is so important to the health of trees that about 80 percent of tree species in the world are involved in these kinds of symbiotic relationships. Because a few species of pine trees in the Northern Hemisphere have so many individuals (think of the acres of boreal forest!), perhaps 99 percent of all individual trees in the world are involved in mycorrhizal relationships.

Mycorrhizal fungi are unable to live without their tree partners, and trees stripped of their fungal partners experience decline or poor growth. Many tree-fungus partnerships are very specific. People have tried for years to cultivate truffles, but the symbiotic partnership between truffles and hazelnut trees limits the success of this venture. Madrone trees are notoriously difficult to grow in nurseries, apparently because culturing the symbiotic mycorrhizal fungi is difficult. On the other hand, some trees are less picky. Bigcone Douglas fir trees in the Transverse Ranges of Southern California may have 50 or more species of mycorrhizal fungal partners attached to a single tree!

Brilliant yellow lichens (*Chrysothrix*) coat rocks along the Klamath River. Photo by Greg de Nevers.

Lichens

Lichens are perhaps the strangest kind of "plant" we will discuss. They may be the organisms that best illustrate the idea of symbiosis. Lichens are not moss, not flowering plants, not fungi, though they have connections to all of these. Lichens are made of two or three very different organisms living entwined together. The lichen body is mostly made of a fungus, with two layers forming a cavity in which an alga or a cyanobacterium lives. The fungus provides a home for the alga and protects it from drying out. The alga is photosynthetic, producing food for itself and for its fungal partner. There are about 1,000 species of lichens in California, each with a different type of fungus, but only two or three kinds of algae and cyanobacteria are involved.

Lichens are often epiphytes, growing on tree trunks or branches. They are very common on rocks, often covering them to the point that you don't see the rock, you just see lichens! They can also live on bare soil. Because their water and mineral nutrients come only from the sky, they are often dry and dormant. It may be that for as much as 90 percent of their lives, lichens are inactive, and for this reason some live amazingly long lives. Some individuals of Arctic species of lichens are estimated to be 11,000 years old, nearly twice as old as the oldest bristlecone pines!

DECOMPOSITION

Every organism on Earth dies. No matter if it is a bacterium that lives a few hours, a turtle that lives 250 years, a bristlecone pine that lives 5,600 years, or a creosote clone in the desert that may live many thousands of years, eventually every individual dies. And when each organism dies, its body is recycled. This is one of the bedrock principles of the biosphere: all matter continually cycles and recycles.

There are lots of animals, bacteria, and fungi that make their living breaking down the undefended bodies of dead organisms. Some attack only freshly dead animals; others specialize in slowly extracting the nutrients from long-dead carcasses. We'll follow the decomposition of a bighorn sheep for a moment to highlight some of the stages decomposition goes through. Imagine a bighorn sheep that falls and dies on impact. Within a few hours, bacteria begin to break into its cells to mine nutrients. Coyotes, foxes, crows, or vultures may rip open and devour the best parts of the carcass (muscles, organs, intestines). Before the scavengers have even finished their meal, flies are on the carcass laying eggs that will hatch into larvae (maggots) that will devour most of the remaining muscle, blood vessels, and tendons. After a few weeks, when the fly maggots have cleaned up all the easily available soft tissues and pupated, there will be a pile of hair, skin, bones, and impressive curved horns left. Even these are concentrated piles of opportunity, and beetles that specialize in hair or skin arrive to deposit eggs. When at last the bones and horns are clean (3 to 6 months), they will be gnawed on by mice seeking calcium. Any or all of these organisms may pee or poop or die as they enjoy their meal. Nearby plants may take up the discarded nutrients that were once the bighorn sheep, became the coyote or the fly, and were discarded to the soil realm.

Decomposition is a complex process and varies depending on the organism that died, the habitat in which it died, the time of year it died, and the scavengers, insects, and microorganisms found where it died. Plants decompose more slowly than animals, but in both cases there is a long line of applicants for the position of decomposer, ready and waiting to take advantage of dead matter. When a leaf falls to the forest floor in the fall, there are specialist insects that eat only the outer layer of the blade, others that rip apart the veins, and still others that lay eggs on the petiole! In the broadest grouping of organisms in an ecosystem, we think of autotrophs and heterotrophs. Autotrophs are the organisms that produce (or capture) their own energy, generally from the sun. In other words, autotrophs make their own food. Heterotrophs are those that eat other organisms to acquire energy. Heterotrophs don't make their own food.

Decomposers are simply a variation on the heterotroph theme, specializing on dead organisms for their energy needs. Many people think of decomposers as including only "microorganisms." Other people also consider fungi to be decomposers. Rather than debating about decomposers, it may be helpful to think of all organisms as either food producers (autotrophs) or food consumers (heterotrophs).

PLANTS AND PEOPLE

People do many things with plants, including move them around. When we move to a new home, we commonly bring plants from the old place. Chinese workers brought tree of heaven (*Ailanthus altissima*) seeds to California to remind them of the lives they left behind in China. Tree of heaven is now widespread in California, with clusters of trees often marking sites of former Chinese habitation. We also import plants of agricultural or horticultural importance. This is quite a two-edged sword. On the one hand, importing avocado plants from Mexico adds a delightful item to our diet and commerce. On the other hand, eucalyptus trees, imported from Australia for ornament and in the mistaken notion that they could be grown for lumber to make railroad ties, have become noxious weeds along the coast of California, where they are long-lived, seed readily, and form large, expanding stands that exclude native California plants.

Walk over to your spice shelf and look at all the choices you are faced with to season your meals: salt, pepper, nutmeg, cinnamon, turmeric, allspice, cloves, and on and on. Now ask yourself which of these spices were available in California before the arrival of Europeans in 1492. Here is the list: bay leaves, bay nuts, and a few roots you probably haven't eaten. Paprika or cayenne may have arrived by trade from Mexico, but in very limited quantities. The point is that the world we take for granted is hugely shaped by the intentional and accidental movement of plants from one continent to another. When we move plants and plant products on such a massive scale, there are inevitably stowaways, unintended hitchhikers who make the journey with the intended contents.

Because people have taken plants with them as they have moved around the world, the world we inhabit today is a composite of elements from diverse regions. Most everyone would agree that it is a good thing that rice from Southeast Asia is now available to grow in the Central Valley of California, which is one of the most productive rice-growing regions in the world. Many people would also be delighted that they can grow rhododendrons from the Himalayas in their yard. What most people don't realize is that transporting plants for agriculture and horticulture can impact native plant and animal communities. For example, transporting cultivated *Rhododendron* plants from Europe was probably the way the water mold *Phytophthora ramorum*, which causes sudden oak death, arrived in California.

We have a series of technical terms to designate the status of plants relative to geography, including *native, nonnative, endemic, paleoendemic, naturalized, invasive*, and *noxious*. Native plants are those thought to be original to a region. Redwoods can be called many names, including endemic, paleoendemic, and native to coastal California and southwestern Oregon. Salt cedar (*Tamarix ramosissima*), on the other hand, is a nonnative. Before people brought it to California as a horticultural plant, salt cedar did not occur in California.

Because *Eucalyptus globulus* and salt cedar are capable of growing, setting seed, and reproducing successfully in California without the help of people, we consider them both nonnative and naturalized. Nonnative plants like these that have successfully become naturalized and have attracted the attention of people as problematic in one way or another are termed invasive plants or invasive exotics. In extreme cases, where naturalized nonnative plants have important economic impacts or negative financial consequences, usually by disrupting agriculture, they reach the ultimate status of being termed noxious weeds.

Climate change is also influencing the distribution of plant communities in California. For example, the downslope edges of forests in the Sierra Nevada have been shown to be retreating to higher elevations. What is unclear is how extreme the changes will be. Changes in species distributions, community composition, and dominant cover types are likely. These vegetation changes will influence the distribution of animals as well as ecosystem processes such as nutrient cycling.

The predicted effects of climate change on plants include

- Shifting of species ranges to the north
- A loss of endemic species with restricted ranges
- An increase in invasive species
- Increases in drought-tolerant species

· Sea level rise, which will alter coastal plant communities, erode beaches, and inundate tidal areas, marshes, and wetlands

A more in-depth discussion of climate change impacts and California's natural communities can be found in the Climate Change section of chapter 7.

Some Important Invasive Species in California

California has nearly 5,900 native plant species, the most of any state in the United States. Over 200 invasive species are threats to California's wildlands and native plant species. Invasive species are second only to development in eliminating native habitat. Some of these species are below.

Arundo donax (arundo, giant reed)

Arundo invades riparian areas, drying up groundwater and displacing native vegetation. Winter rains can distribute reproductive parts of this plant down the creek.

Tamarix (tamarisk, salt cedar)

Tamarisk invasions have numerous impacts, including clogging streams, drying up wetlands, and increasing the salinity of soil. About 1.5 million acres in the US Southwest have been invaded by tamarisk.

Rubus armeniacus (Himalayan blackberry)

Blackberries are loved by pie makers, but they have insidious effects on the wetlands they invade.

Genista monspessulana (French broom)

French broom is highly destructive to native plants and can cover whole hillsides. Despite this, it is still sold as an ornamental in gardening stores.

Centaurea solstitialis L. (yellow star thistle)

Yellow star thistle can be found in over 15 million acres in California, including rangelands, native grasslands, orchards, vineyards, pastures, roadsides, and wasteland areas.

Aegilops (goat grass)

Goat grass is of concern because it has the dual ability to hybridize with winter wheat and to invade serpentine soils. Invasion of serpentine soils is a concern, as those soils also harbor many species endemic to California.

Native American Plant Uses

Indigenous Peoples of California knew the regional flora so well that they were able to make their living from the land without importing agricultural crops from other regions. They knew which plants and animals were good to eat and how to tend the land and some plants to provide good returns for food and fiber. Land managers and conservationists recognize that Indigenous knowledge and stewardship methods have the potential to improve ecosystem resilience, protect biodiversity, and benefit society. Advancing Tribal community leadership and the practice of Indigenous stewardship will help us restore California's precious natural and cultural heritage.

There is abundant evidence that Native Californians harvested and stored great quantities of acorns, processed them into staple food items, and revered them in their religious worldview. Native Peoples did not need to plant oak trees, but they "tended" them to improve the acorn crop. One of the forms of TLC Native Californians applied to the oak trees they intended to harvest was to burn around and beneath them! Fire was a primary "agricultural" tool in pre-European California. Broadcast burning was practiced on a grand scale, and probably for multiple purposes. One of the purposes of repeated burning, in many places on an annual basis, was to reduce or eliminate shrubs from the oak woodland understory. This made travel easier, facilitated hunting by opening up visual corridors and providing food for elk and deer, and improved the collection of acorns by clearing the ground the acorns would fall on and eliminating pests that attack acorns. More broadly, repeated burning opened meadows and increased their size. Grasslands were another of the resource-harvest areas Native Californians relied on, for digging liliaceous bulbs and for harvesting the seeds of annuals and herbaceous perennials. Perhaps the second most important plant food resource in California was pinole, a mixture of the seeds of a variety of herbaceous plants collected by beating seed heads into baskets.

There is a shrub common in coastal California called Indian plum, or oso berry (*Oemleria cerasiformis*). The fruits of this shrub were highly sought after by many Tribal communities in California. Oso berry plums are small, with a single, large seed. The taste is initially sweet, but there is a strong, bitter aftertaste. The fact that these bitter plums were considered a delicacy is a testament to the narrow range of fruit choices available to pre-Columbian Californians.

Another layer of interdependence Indigenous Peoples had with plants was their use of plants for tools and fibers. The baskets used to collect and store both pinole and acorns were made from various plant fibers, including willow, sedge, iris, hazelnut, ninebark (*Physocarpus*), and redbud. Basket making in California reached levels of technical skill unsurpassed anywhere on Earth. Baskets were woven that were watertight! Basket making also intersected the realm of the mystical when designs for a basket were found in a weaver's dreams. Digging sticks for harvesting liliaceous bulbs were crafted of mountain mahogany and toyon. Arrows were made of the straight stems of oceanspray (*Holodiscus discolor*). Houses and temporary shelters were made from a wide variety of plants, depending on the species available at a particular site.

We continue to depend on and interact with plants in numerous and sometimes surprising ways. California has been transformed since the arrival of Euro-Americans, with the damming of rivers for irrigation and flood control, the removal of oak woodlands, the plowing of the Central Valley for agriculture, the elimination of elk and pronghorn, and extensive cattle grazing being perhaps the most noticeable. We have imported an astonishing array of plants to California from all corners of the world. California is a plant enthusiast's dream come true. The variety of plants available for agriculture, horticulture, crafts, and study is truly remarkable.

Explore!

TAKE A CREEK WALK

Walk a local creek. Can you identify the riparian plants? (Beware of poison oak!!) Can you tell where the riparian community ends and the upland community begins? If you don't want to walk alone, contact a local park or preserve, a creek association, the California Native Plant Society, or a Sierra Club or Audubon chapter for information about creek walks.

LEARN ABOUT A LOCAL PLANT COMMUNITY

Choose a piece of land you care about, such as your own property, a local oak woodland, a redwood grove, a creek, a wetland, or a local park. Spend a day walking and sketching it and writing down what you notice. Then learn its history. Your local library, the county clerk's office, and the park or preserve's administration offices will have historical records, but you may find that the best source of information is oral history. Ask someone who has lived in the area for a long time about that land.

(continued)

VOLUNTEER AT A NATIVE PLANT NURSERY

This is a fun way to learn local plants and also learn about local plant issues in your community. Ask at conventional nurseries, or call your local California Native Plant Society chapter to find a native plant nursery.

EXPLORE YOUR NEIGHBORHOOD

Spend a half hour in your neighborhood without any interruptions. Take a journal and write down and/or draw what you see, hear, smell. Survey the plants and make note of any you can't identify. Pick three to learn more about. Then go to a local park and see if you can find any plants that are the same as the ones in your yard.

PARTICIPATE IN A RESTORATION PROJECT

Restoration projects will get you pulling weeds, planting native plants, and if you are lucky, collecting seeds. They are fun and a great way to increase your knowledge of local plants. Check with local parks and creek "friends of" associations to see if they hold regular restoration days.

PRESS PLANTS

Pressing plants is another great way to get to know local plants. To make a small field plant press, cut two 8½-inch by 11-inch pieces of cardboard, and place folded newspapers between the cardboard. Collect the stem with attached leaves and, if at all possible, flowers and/or fruits, place them in the folded sheet of newspaper, sandwich them between the two sheets of cardboard, and loosely tie the bundle closed with string.

**LANDSCAPE WITH NATIVE PLANTS AND
WILDFLOWERS IN YOUR YARD**

About 30 to 60 percent of urban fresh water is used for maintaining US lawns, along with the application of 67 million pounds of synthetic pesticides. Native landscaping may help reduce your water use, eliminate the need for pesticides, and make your yard unique and attractive to native fauna such as butterflies.

MAKE ART ABOUT PLANTS

Sit in a natural place you love and write a poem or a song, draw a picture, make a painting, or take a photograph of something that inspires you. Try to incorporate naturalist details into your work, such as the shape of the leaves, pollinators, insects that use the plants, seeds, or whatever you notice and feel drawn to.

5

Shrublands, Woodlands, and Forests

Shrublands, woodlands, and forests are places of beauty and refuge, often inspiring awe and a connection to something greater than ourselves. They provide habitat for birds, insects, fish, amphibians, and reptiles. They moderate temperature and streamflows. They sequester carbon. They are the source of wood, a hugely important natural resource that we depend on for lumber, paper, and fuel. For these reasons, feelings about shrublands, woodlands, and forests run deep.

Shrublands, woodlands, and forests are areas with woody vegetation, of many kinds. California is notable for the diversity of its plants, animals, and habitats, and much of terrestrial California is covered in some kind of woody vegetation. Why is this?

Think of the bare ground left behind by a retreating glacier or a volcanic eruption. What becomes of that land? Plants quickly occupy the opening, in a process called primary succession. Now ask yourself, Why do some spots have grasslands, but other places have shrubs or trees? Most places on Earth will succeed to woody vegetation unless there is some kind of intervention. Buffalo are the classic intervention in grasslands, stomping, trampling, and grazing huge areas and excluding woody vegetation in the process. For millions of years huge herds of browsers and grazers maintained vast areas of North America as grasslands. Fire also maintains grasslands. If fires burn often enough, all woody plants are excluded. Prior to the arrival of people in North America 18,000 years ago, fires started by landslides (when rocks rubbed together and created sparks), volcanic eruptions, and lightning shaped the vegetation of California.

With the arrival of humans in California, most of the large grazers were eliminated, and a new source of fire was introduced. People use fire for multiple purposes, shaping the vegetation of California in the process. When you look at some natural or seminatural area, you should ask yourself, Why is this area a grassland? Or, Why is this area a shrubland? Or, Why is this area a forest?

California has nearly 33 million acres of forests—about a third of the state's total acreage. Shrublands, including chaparral in cismontane California and the desert shrublands east of the mountains, cover perhaps 37 million acres. Oak woodlands in California may cover 10 million acres. In California there are diverse kinds of woodlands, forests, and shrublands: those primarily covered with conifers (trees bearing cones) such as redwoods, Douglas firs, and pines; those primarily covered with broadleaf hardwoods, including oaks, madrones, maples, and sycamores; riparian forests of deciduous trees; and desert woodlands with creosote bushes, Joshua trees, pinyon pines, and yuccas. All of these habitat types provide the state's residents with vital services, including wildlife habitat, water storage, wood products, recreational opportunities, and scenery. Another increasingly recognized value of woody vegetation is the ability to store carbon, helping to offset the adverse impacts of people releasing increasing amounts of carbon dioxide and other greenhouse gases into the atmosphere.

Coniferous and broadleaf forests are very different. In the Sierra Nevada and Southern California mountains, conifer forests generally occur at higher elevations than oak woodlands. The areas where conifers grow are often colder and wetter than in hardwood forests. Conifers are mostly evergreen, and broadleaf trees are often deciduous, so the amounts and seasonality of light that can penetrate the two types of canopies are very different, especially after the hardwoods' leaves have dropped. Conifer stands can be dense, with only diffuse sunlight penetrating to the ground. The trees and shrubs in the understory are diverse: rhododendrons, dogwoods, and huckleberries may overtop a hiker. Conifers generally grow faster than hardwoods, and the wood products produced, including lumber and paper, are more valuable. The trees in conifer forests can grow to enormous size. Coastal redwoods, for instance, can be over 300 feet tall, with a diameter of over 20 feet.

At lower elevation in the foothills and the Central Valley, oaks often dominate the landscape. These irregularly shaped trees may form dense

A diverse shrubland in the Kingston Range of the Mojave Desert. Photo by Greg de Nevers.

stands, be widely spaced as in a savanna, or stand as isolated individuals. Oaks provide valuable habitat for animals and are the primary food resource for deer, bears, wood ducks, acorn woodpeckers, and many other animals. Acorns from oak trees were a key food resource that could be stored year-round and, along with salmon and other protein sources, allowed the human population of California to be more dense than anywhere else in North America prior to first European contact. Imagine, a nonagricultural people with a higher population density than the corn-farming mound builders of the Mississippi or the Aztecs!

Throughout California, vast areas are covered in shrubs and small trees. In cismontane California, these areas are often called chaparral, and in the deserts, they may be called desert shrublands. These shrublands are one of the most characteristic elements of the California landscape. Chaparral is often dominated by manzanita (*Arctostaphylos*) and wild lilac (*Ceanothus*). These two genera have multiple unique species scattered throughout the state, with many local endemics. In the desert, there are vast areas covered with creosote bushes—very diffuse, strongly aromatic shrubs 6 to 10 feet tall. Creosote clones may live for thousands of years! The Los Angeles basin and the deserts also harbor

Blue oak savanna during the rainy season, when densely spaced blue oaks have all leafed out. Photo courtesy of Greg Damron.

a tremendous diversity of woody plants, with multiple distinct habitats and many kinds of small trees and shrubs.

HISTORY OF CALIFORNIA FORESTS AND THEIR STEWARDSHIP

For at least 18,000 years people have lived in California. Indigenous stewardship has shaped the land to suit social and ecological needs, with fire used as the primary tool. Native Californians burned around villages to protect their homes from wildfire, to encourage seed-producing plants they harvested for food, and to stimulate bulb-producing plants they dug to eat. Farther from village sites, Native Californians burned vast landscapes to alter the vegetation to their advantage. The landscape Europeans encountered when they first entered California was an intimately stewarded, completely occupied, highly useful life zone.

When Europeans first appropriated California from its earlier inhabitants in the mid-1800s, the forests, woodlands, and shrublands they captured were vast. Early European colonists saw mixed conifer forests in the Sierra Nevada, coastal forests that contained massive Douglas firs, and awe-inspiring redwoods, and they imagined how they could use these huge trees when they were cut. During this time, wood

Oak Woodlands in Peril

Though the majority of Californians live in close proximity to oak woodlands, few people realize the ecological and economic importance of oak woodland habitat or the frightening rate at which it is disappearing. Many factors threaten the integrity of oak woodlands, but the number one threat remains development: housing development, agricultural and vineyard conversion, and tree cutting for firewood and forage. These types of land use lead to high rates of habitat loss and fragmentation that are likely to continue. It is estimated that a third of California's oak woodlands and forests have been lost to agriculture and urban development, and 750,000 more acres are expected to be lost to development within the next 25 years. The oak woodlands of the Central Valley and Sierra Foothills face the most immediate threats. Land use planning generally takes place at a local or regional level, where it can be affected by local political and financial influences. Conservation of oak woodlands, therefore, presents a challenge to regional planners and elected officials to create solutions to complex ecological and social problems.

products were generally taken by removing the biggest and straightest trees.

The available technology largely dictated how and where harvesting took place. In the early years of European settlement, water was the most useful method of transporting logs to mills and boards to markets, so logging was concentrated near rivers and on the Pacific Ocean. The first time San Francisco was built, between 1850 and 1906, much of the lumber came by ship from Puget Sound! In the Sierra Nevada, trees were hauled to streams, and "splash dams" were built to collect sufficient water to carry trees downstream. The torrents of water rushing from opened splash dams stripped water courses of boulders, spawning gravels (used by salmon and steelhead), large woody debris, and riparian trees. These floods widened streams and caused massive erosion. The effects of the rush of water and logs downstream can still be seen in the form of scoured streambeds in the mountains and sediment deposits in valleys. Even today, some of these streams have low habitat complexity and support fewer juvenile salmonids as a result.

When railroads became available to transport raw materials, harvesting took place where access was easy and the rail transport cheap.

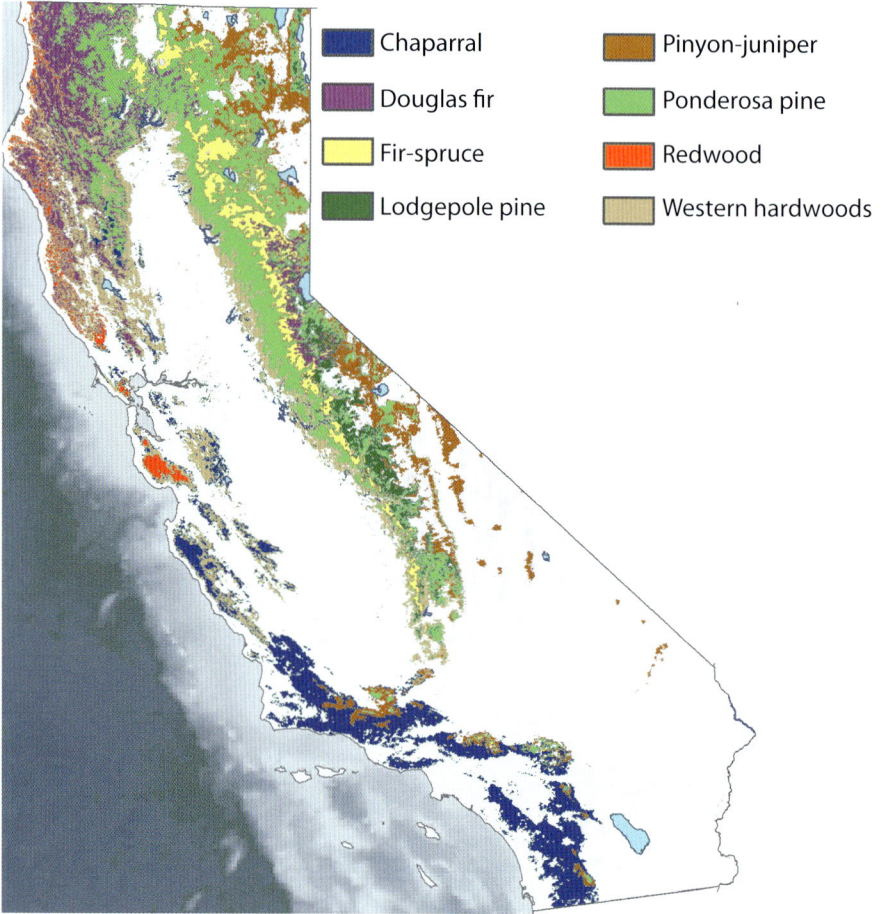

California forest cover types. The white areas are grasslands or desert shrublands. Adapted from the US Geological Survey.

It was often easiest to take all of the trees at these places, so clear-cutting—a harvesting system that removes all of the trees on a tract of land—became common. This was not an intended forest management practice. It was just the most expedient method of tree removal. After clear-cutting, people relied on natural regeneration to restock the harvested areas, but few people were considering whole-system impacts.

By the beginning of the twentieth century, forests no longer seemed inexhaustible. The European settlers began to encounter the era of limits. Economists began to have concerns about wood supplies, and ecologists worried about watershed protection and impacts to nontarget species. Huge wildfires occurred during this period, which led to rules for treating slash (wood left behind after tree cutting) in order to protect the increasingly rare, and thus valuable, old-growth timber stock. As a result of these realizations, Euro-Americans put forward the idea that the US government should get involved in forest protection and management. Under the leadership of Gifford Pinchot, the US Forest Service was established by Congress in 1905, Pinchot was made its first chief, and a network of national forests was established. The number expanded rapidly, and compared with other regions of the country, California had extensive national forests designated. Forest Service lands were opened for public use, including timber harvesting, grazing, mining, and recreation.

By the early twentieth century, many people felt that good stewardship was essential and hoped that government ownership would help protect forests into the future. Pinchot recognized that forests provided multiple benefits and championed wise use of resources. In addition to promoting multiple use, Pinchot felt that forestry should be evidence based and that the goal for determining successful management was "the greatest good for the greatest number of people for the longest time." Under his leadership, some of the worst practices, in terms of adverse impacts on the environment (for example, hydraulic mining), were outlawed on national forest lands. Pinchot's multiple-use philosophy brought him into conflict with those more interested in exploiting natural resources for profit, as well as with those, like John Muir, who advocated complete protection for some areas, excluding any of the extractive uses. Muir felt that publicly owned lands should not be used for resource exploitation and that people in extractive industries, such as mining and lumber, should look elsewhere for their products. Muir's writings and lobbying were influential in the creation of the National Park system. The conservation and preservation principles championed by Pinchot and Muir have influenced natural resource management for generations.

After the Second World War, demand for wood products increased dramatically and harvesting skyrocketed, fueled largely by increases in home construction. From 1950 to 1966, for instance, twice as much timber was cut as in the previous 45 years. Both private timberlands and public lands managed by the US Forest Service and the Bureau of Land Management were heavily harvested to provide wood products. The timber owners and managers built roads into previously roadless areas and developed harvest practices that included the clear-cutting of old-growth forests and the spraying of herbicides to suppress noncommercial

Gifford Pinchot (1865–1946). Photo courtesy of the US Forest Service, Grey Towers National Historic Site.

trees, like tanbark oaks. Unfortunately, some people failed to notice that this rate of harvest was unsustainable.

During this period, not only conifer forests, but also oak woodlands and chaparral were impacted by human activities. Between 1945 and 1970, approximately 2 million acres of woodlands and chaparral in California were cleared of woody vegetation to create grazing land. While clearing sometimes did result in forage increases, the forage increases often didn't last, while the adverse impacts—including soil erosion, degraded wildlife habitat, and treeless landscapes—persisted. Fortunately, such clearing for range use is slowing. But in the last two decades there has been an increase in conversion of oak woodlands and chaparral to intensive agriculture, such as vineyards, and for suburban and rural residential development.

Between 1890 and 1910, a series of catastrophic wildfires in the western United States killed thousands of people and destroyed huge quantities of valuable timber. The Forest Service responded to these fires by instituting a policy of total fire suppression. This was formalized in 1933 with the 10 a.m. rule. The Forest Service aimed to put out every wildfire by 10 a.m. the day after it was spotted. Forest Service managers imagined a world with no fire, thus no loss of commercially valuable timber nor destroyed towns and loss of life. Unfortunately, the idea of forests with no fire is a mirage. For many western forests there are basi-

John Muir (1838–1914)

John Muir. Photo courtesy of the National Park Service.

The clearest way into the Universe is through a forest wilderness.

—John Muir

John Muir was a Scottish-born naturalist and writer. He is best remembered for advocating that everyone should experience wilderness, and fighting for its preservation. He was among the first to call

(continued)

John Muir (1838–1914) *(continued)*

attention to glaciers in the Sierra Nevada and their role in carving Yosemite Valley. His writings and lobbying were influential in the creation of the very idea of national parks. Muir was a cofounder of the Sierra Club and figured prominently in many of the important environmental controversies of his day, including the question of whether to put a dam inside Yosemite National Park to supply San Francisco with water. Muir seems not to have recognized the role of Native American land stewardship in the "wilderness" he saw in Yosemite, and he has been criticized for advocating a human-free park. Some of his journals include disparaging comments about Native Americans and Black Americans, while his later remarks reflect greater respect for Indigenous Peoples in California. Muir and Gifford Pinchot were initially friends but became rivals when their philosophical differences became clear, with Pinchot supporting "multiple uses" of national forests and Muir favoring wilderness without commercial use. Muir's point of view became known as preservation, while Pinchot's became known as conservation.

cally two choices: frequent, often low-intensity fires, or infrequent high-intensity fires. The policy of total fire suppression was opposed by Tribal communities and ranchers, who recognized the utility of fire in land stewardship and the futility of the myth of no fire. However, the Forest Service persisted in this well-intentioned but near-sighted policy.

The forests we have inherited today are strongly shaped by the policy of fire suppression. Notable outcomes are forests that are far too dense, sometimes called dog-hair thickets, that are prone to high-intensity canopy fires. Fires in these types of forests under severe weather conditions (dry, hot, with high wind) are essentially impossible to control. One of the huge questions facing land stewards today is how to mitigate the results of the policy of fire suppression. Most everyone agrees that the goal is fire-resistant and fire-resilient forests, but reinstating fire in the overstocked forests of today is a challenge.

In the 1960s the environmental movement grew rapidly, and the public became increasingly aware of the way shrublands, forests, and woodlands were being mismanaged. Environmentalists opposed the clear-cutting of conifers and the conversion of oak woodlands, and Tribes were vocal opponents of herbicide spraying and fire suppression.

While clear-cutting was often the most expedient harvest method, it adversely affected human health, fisheries, and wildlife habitat and was responsible for increases in erosion and sedimentation. There was also concern about how commercially oriented forest management practices were affecting the overall environment.

Rachel Carson's book *Silent Spring* awakened the public to the dangers of pesticide use, and people became concerned that herbicides sprayed in forests to kill noncommercial vegetation could get into waterways and impact wildlife and human health. Environmentalists also became concerned that old-growth stands were being cut at an alarming rate and that old-growth-dependent species like the fisher, the marbled murrelet, and the spotted owl would disappear with the forests they depend on. There was an emerging understanding that forests, woodlands, and shrublands are more than just assemblages of shrubs and trees or supplies of timber, but also storehouses of biological diversity. All woody landscapes are now recognized as special places that provide inspiration and renewal. For all these reasons, efforts were begun to protect shrublands, woodlands, and forests.

The precursor to the Endangered Species Act (ESA) was passed by Congress in 1966, and the ESA was passed in 1973. It was aimed at protecting species that were threatened with extinction, and it provided a vehicle for opponents of the prevailing forest management practices to challenge them in court. Since some wildlife species, such as the northern spotted owl and the marbled murrelet, require large old trees for hunting and breeding, the courts ruled that existing forest management practices were insufficient to guarantee these species' survival. This ruling and other legal challenges helped to trigger changes that led to new and different approaches to forest management.

The impacts on wildlife became a frequent determining factor in timber harvest decisions, and ultimately there were fewer and smaller harvest areas and more retention of old-growth trees. However, many people felt that not enough had changed. Some individuals sat or even camped out in trees to protect them from being cut down, and others tried to stop logging on private lands by physically blocking vehicles or interfering in other ways. In addition, legal challenges to proposed logging plans were vigorous and often successful. Such challenges led to a transformation in the way forests are managed. On federal lands, ecosystem management became public policy. The underlying principles of ecosystem management are that biodiversity and ecosystem health should be managed at the landscape level, not the stand level; processes

Working Together: The Redwood Forest Foundation

The Redwood Forest Foundation is an example of how people with different backgrounds and interests can come together for a common goal and can change how local forest resources are managed. For decades, the communities in the redwood region along California's North Coast experienced conflict and argument over the logging of old-growth forests. The conflict was partly due to the fact that decisions affecting local communities were often made by absentee, corporate landowners. In 1997, a nonprofit organization calling itself the Redwood Forest Foundation, Inc. (RFFI) was formed. Its stated mission was "to acquire, protect, restore, and manage forestlands and other related resources in the Redwood Region for the long-term benefit of the communities located there." It began by building a diverse board of directors, including people who had often been adversaries in forest conflicts. The directors included representatives from the timber industry (a mill owner and Registered Professional Foresters), community activist groups (Earth First! and the Sierra Club), the banking community (a stockbroker, a banker), and the University of California Cooperative Extension. They set out to create a new structure of ownership and community partnering unprecedented in the region. RFFI established county-based advisory committees that played the traditional "shareholder" roles in its decision-making process. These committees also reflected the many talents and diverse experiences of the local communities.

In June 2007, 10 years after its formation, RFFI acquired nearly 51,000 acres of coastal redwood lands in northwestern Mendocino County, with $65 million in financing from Bank of America. The bank's enthusiasm for working with RFFI was due largely to the community-oriented structure of RFFI's board and the credibility of its members and advisors. The structure of the loan included provisions to protect the forests and prevent overharvesting of timber in order to reduce debt. "This is the beginning of a new era for our local community," said Art Harwood, then-president of RFFI. "We are banding together to protect and manage our forests. We are pulling together private capital, and the hopes and aspirations of people from all walks of life to create a bright beacon for our future. We are doing this by ending the 30 years of fighting, and focusing on what unites us."

The RFFI illustrates the idea that forest stewardship in California is changing. In the past, forest managers often focused almost exclusively on timber production. Today, new organizations such as the RFFI realize that cooperation among competing interests is often crucial if they want to steward forests to provide critical habitat, increase biodiversity, and improve regional economic vitality.

Contributed by Greg Giusti, Forest Advisor-Emeritus, UC Cooperative Extension.

such as nutrient and water cycling are critical to take into account; and the emphasis should be on the ecosystem as a whole, not only on products such as timber or on individual wildlife species. Ecosystem management also takes into account the fact that people are part of the equation and that management should consider the needs of the local community, as well as being ecologically based.

Today, the debate about how to manage forests in California continues. The Forest Service and Bureau of Land Management are now both dedicated to practicing ecosystem-based management. As a result, harvesting has been greatly reduced on public lands, and areas of old-growth and mature (late-succession) forests have been set aside. Laws have been passed that regulate timber harvest on federal, state, and private lands. This has resulted in an increased emphasis on "uneven-aged management" and an attempt to make forests more diverse, with a greater range in age and size classes and species, as well as adequate dead material, both as standing snags and as dead and down wood. Forested areas are also recognized as habitat corridors that wide-ranging wildlife species use to travel between habitat patches. The difference between commercial timber plantations, which should not be called "forests," and national forests, which are managed for overall ecosystem values, grows ever wider.

In the last several decades, there have been dramatic changes in the management of oak woodlands as well. In the early 1980s, there were calls for increased state regulation and for rules that would limit the cutting of oaks and other hardwood species. However, the California Board of Forestry and Fire Protection was reluctant to adopt statewide regulation and instead passed on to the counties the responsibility to develop conservation practices to protect oak woodland resources. The result is a patchwork of local policies. Many California jurisdictions have no policy at all, and the ones that exist have been applied with mixed success. Woodland habitat is still being lost as oak woodland is converted to agricultural land and used for housing and commercial development.

Similarly, in the Mojave and Colorado Deserts large tracts of shrubland have been denuded for solar power installations. The lack of public awareness of the biodiversity and ecological richness of desert ecosystems underpins these conversions.

From a social standpoint, public land managers now recognize that they must consider the effects of forest management on people and communities and that the public should have input into the decisions that affect them. There are also efforts to operate more at the landscape level and to bring together public and private landowners and stewards for cooperative decision-making. Numerous watershed protection groups, in shrublands, woodlands, and conifer forests, have been established to address mutual problems and develop innovative solutions (see Collaborative Conservation in chapter 8). There is a general belief that land stewardship should be flexible and adaptive, not rigid, and management practices should be modified if something doesn't work or if new information or experience shows us a better way.

LANDSCAPE DYNAMICS

Shrublands, woodlands, and forests are dynamic habitats. Through both human and natural disturbances, these plant communities are in continual flux. The composition of species in an area changes, depending on the age of the community and the time span between disturbances, with smaller-magnitude disturbances having greater impacts as the stand matures. For example, all else being equal, a less dense coniferous forest that had recently burned wouldn't have enough fuel to support a high-intensity fire, so it would do less damage than a fire in a more dense forest that hadn't seen recent human or natural disturbances.

As woody plants grow and develop, they change the environment. On a site that has not previously borne vegetation, such as a recent lava flow or an exposed slope after a landslide, plants may be exposed to full sunlight. Initially, only plants that can tolerate high light and low nutrient levels become established. However, as light-loving plants grow, they provide shade. Eventually a different group of plants begins to grow in the understory—those that can tolerate high levels of shade. Other changes in the environment can also affect community composition. For instance, some sites are initially deficient in nitrogen. A few pioneer species (the first ones to grow) have nodules on their roots that let them fix nitrogen, that is, convert nitrogen in the air to a form that plants can use. California examples include alder (*Alnus*) and buckbrush (*Ceanothus*). Eventually the fixed nitrogen increases the fertility of the soil, and other plants begin to grow there. While plants affect each other and the available resources, it is often the frequency and duration of the disturbances in any given habitat that dictate the community composition.

There have been many studies on succession to try to predict changes in the composition or structure of communities following disturbances such as logging and fire. These studies have found that it can be hard to generalize the path of change. There are many site and landscape characteristics that influence a community's response to disturbance.

Plant communities change slowly or rapidly, depending on the disturbances that occur. As a budding naturalist, take time to notice the size of the trees in a stand, as well as what makes up the understory, and think about how past disturbances such as harvesting, grazing, or fire may have influenced what you see.

Forest change can also occur because of disease. When disease organisms are moved from continent to continent by people, they can dramatically alter ecosystems. Chestnut blight, for example, was a fungus accidentally imported on nursery trees to the eastern United States around 1900. In about 40 years, it killed almost all of the chestnuts in the eastern deciduous forests, forever changing the look, feel, and function of those forests. Chestnuts used to be the largest and most abundant trees in many eastern forests, provided the keystone food resource for many animals, and grew to a diameter of 14 feet! Now they are gone, and the forests are dramatically altered.

We are currently in the midst of what may be California's version of chestnut blight. A fungus, probably imported accidentally on cultivated rhododendrons from Europe, is dramatically changing the structure of coastal oak woodlands by killing thousands of tan oaks and coast live oaks. Will our grandchildren grow up without ever smelling the distinctive tang of blooming tan oaks in June? On California mountaintops the combination of white pine blister rust (accidentally introduced by foresters), native mountain pine beetles, altered fire regimes, and climate change is devastating western North American stands of whitebark pine (*Pinus albicaulis*). The tree's range is expected to shrink by 70 percent within two decades, and in 2022 the US Fish and Wildlife Service listed the species as endangered.

CALIFORNIANS AND WILDFIRE

Fires have always occurred in California and have been instrumental in shaping our landscape. Lightning, landslides, and volcanoes ignited fires for millions of years before people came to North America. For the past 18,000 years fires were intentionally set by Native Americans. There was a widespread recognition by Native Americans that regular burning was beneficial for a variety of reasons: it improved forest health, made hunting easier, stimulated the growth of plants used in basket weaving, made it easier to collect acorns and other products (seeds, mushrooms, berries),

Sudden Oak Death and Threats to Oaks

Sudden oak death (SOD) is a disease that infects a large number of California plants and can be lethal for several native California oak species. It is caused by *Phytophthora ramorum*, a fungus-like water mold that causes oozing sores that can girdle and kill mature trees. SOD is native to the mountains of East and Southeast Asia and was introduced to North America and Europe through infected ornamental plants. Since its appearance in California in 1995, SOD has killed tens of millions of tanbark oak, coast live oak, California black oak, Shreve's oak, and canyon live oak trees. The pathogen does not reproduce on oaks but on California bay laurel and tanoak leaves. SOD spreads through the movement of live infected plants and through the movement of infected green waste. SOD has been confirmed in 15 coastal counties, extending from Monterey to Del Norte County, but it appears that conditions farther inland are too hot and dry to permit its spread there. The participatory science program called SOD Blitz empowers concerned people to map the location of the disease and to help identify those hotspots that require preventive actions to preserve oaks.

Another nonnative threat, the goldspotted oak borer (GSOB, *Agrilus auroguttatus*, referred to as the golden SOB!!) has been killing thousands of oaks in the mountains of San Diego County. It has now spread to adjacent counties. The GSOB appears to have come from Southeast Asia, like SOD. To date, it has attacked coast live oaks, California black oaks, and canyon live oaks. The golden SOB attacks large, vigorous, healthy trees, including urban trees in people's yards. Tree mortality is enhanced by the pathogen *Fusarium*, which is introduced by the GSOB when it bores into trees.

Perhaps more important than the GSOB is the introduction of the emerald ash borer (*Agrilus planipennis*), another wood-boring beetle from Asia. As this potent ash specialist has spread across North America since 2002, it has killed up to 99 percent of the ash trees in its path! It has the potential to dramatically alter the riparian forests of California by killing most ash trees and impacting the associated species that depend on ash trees, for instance lichens.

Moving firewood is an important method of spreading several invasive species, including the golden SOB and the emerald ash borer. Federal, state, and local resource managers are working to raise public awareness as a part of their effort to prevent the spread of pest species through the movement of firewood. Do your part and don't move firewood!

Mechoopda Tribal and community members conduct a cultural burn to promote *osoko sawi* (deergrass, *Muhlenbergia rigens*) near Chico, CA. Photo by Ali Meders-Knight.

and reduced the likelihood of catastrophic fires. Early Euro-American ranchers regularly burned their lands—especially oak woodlands—to improve forage production. But about 100 years ago, fire management practices in California dramatically changed, and the era of fire suppression began. This largely resulted from dismay at the loss of human lives, homes, and timber resources in massive wildfires near the beginning of the twentieth century. It was reasoned that if wildfires could be put out soon after they started, they would be contained and losses would be reduced. Firefighters were very successful at their job, and as a result, fire as a natural ecosystem process was dramatically reduced.

Many decades passed before people saw a downside to such effective fire suppression. In the past, frequent low-intensity fires had opened up forests and removed more shrubs and dead wood, leaving fewer, larger, fire-resistant trees. The suppression of fires during the past 100 years caused dramatic ecosystem changes. In low- and middle-elevation conifer forests, shade-tolerant species such as Douglas fir and true firs (*Abies*), previously eliminated by regular burning, became much more common. Shade-intolerant species such as ponderosa pine were crowded out. Dead material on the forest floor built up, and ladder fuels—those fuels that allow a fire to move from the ground to the tops of trees—increased. Similar, though not as dramatic, changes occurred in oak woodlands as dead material accumulated in the understory and shrub species increased in some areas. In some low-elevation forests, especially in coastal foothills, the removal of fire caused a conversion from savanna-like woodlands to forests dominated by conifers, especially Douglas fir. A consequence of these changes is that when fires do start in California forests—especially during periods of low humidity, high temperatures, and strong winds—they are much more difficult to control, and they tend to burn entire forests rather than thinning the forests. Some of these fires have become catastrophic stand-destroying fires that have burned large, established trees as well as homes and people.

Unlike many forests and woodlands, chaparral is adapted to infrequent, high-intensity, stand-replacing fire. The shrubs and herbs of this community love fire, but only every 25 to 100 years. Many shrubs of the chaparral reproduce mostly after stand-destroying fires, and the perennial herbs that have been suppressed by the chaparral canopy grow luxuriantly and seed prolifically for a few years, until the chaparral shrubs reestablish. In Southern California chaparral, too-frequent fire, mostly based on human ignitions, is the new inappropriate norm.

California's deserts are also experiencing fires more frequently and with greater severity than in the past. The spread of invasive plants like cheatgrass and red brome has significantly altered the desert fire landscape by connecting widely spaced shrubs with continuous fine fuels. As in the chaparral, human ignitions have increased the frequency of fires in the deserts. This increased frequency and homogeneous spread of wildfires poses a threat to the iconic Joshua trees, creosote bushes, and ocotillos, as well as to the pinyon pines that are an ecocultural food source for Native Peoples. Focused efforts at weed control, fire suppression, and protection of climate refugia are all helping to conserve these fragile desert ecosystems.

Today it is recognized that the structural changes in forests need to be reversed if we are to reduce catastrophic fires. Managed fire, canopy thinning, and removal of understory plants are tools to make forests more fire-resistant. Regulatory hurdles, as well as the scale of the problem, have thus far prevented these treatments from achieving statewide success. The legacy of our current spate of megafires is a continual spiral of forest alteration to conditions that are difficult to bump toward sustainability. Perhaps even more important is the legacy of devastated lives

Animal and plant movement between remaining fragments is impeded across this urban and agricultural landscape, Sonoma County. Vineyards, orchards, pastures, reservoirs, and roads penetrate and replace oak woodlands and chaparral, and the flatter valley margins are in industrial and residential land uses. Photo by Adina Merenlender.

and traumatized survivors each of these events produces. It will take many perspectives and innovative solutions to rebalance our urban, suburban, and rural lives with the fire-prone forests we have engineered. One hopeful sign is the many counties that have passed building codes that require using fireproof materials and designs when doing construction in the wildland-urban interface. It may also take us rethinking our desire to live in remote, rural areas surrounded by beautiful trees and expecting society to protect us from an unstoppable force of nature: fire!

FRAGMENTATION OF FORESTS

Another recent trend impacting California's forests is fragmentation, as large tracts of land are sold and subdivided. Some of these lands continue to be managed for natural resource production, while others are converted to home sites with a range of factors that were previously absent: roads, buildings, power lines, noise, and pets. These changes alter the character of the landscape, increase water runoff, alter stream courses, adversely affect wildlife, alter scenery, and impair ecosystem processes. The trend toward fragmentation is especially evident in chaparral and oak woodlands as people move from densely populated urban areas to seek the beauty and solitude that chaparral and woodlands provide. While development pressures are less intense in coniferous forests, they still exist and bring with them a range of problems. In addition to impacting wildlife species that need protected, interior habitat, fragmentation in coniferous forests can reduce management options. For instance, it becomes increasingly difficult to protect people and homes from wildfire if houses are scattered throughout the forest.

Private land conservation tools can provide incentives to prevent habitat loss and fragmentation while maintaining compatible land use practices and revenue from compatible activities such as sustainable forestry or livestock production. One such commonly used tool is the conservation easement, an incentive-based approach that relies on continued private ownership and management of land with an easement attached to the title that limits some development or other activities on the land. Easements typically cost less than fully acquiring the land for protection purposes and, perhaps more important, leave the management costs and opportunities for other activities on the land to the private landowner while at the same time constraining development.

Zoning laws that require large minimum parcel sizes in wildland areas also help protect those lands. But as more Californians seek to live in rural areas, there is greater pressure to change zoning and to fragment the wildlands. In Nevada County, for example, the median size of landholdings in 1957 was 551 acres, but by 2001 it was only 8.9 acres, and today many of the parcels are zoned to accommodate 5-acre lots. The pressure to change zoning is particularly intense where large sums of money can be made by the sale or development of land. Improved land use planning, acquisition of conservation easements, and other private land-conservation tools are needed if large forest and woodland parcels are to be protected, allowing them to continue to provide critical goods and services to the people of California.

CARBON SEQUESTRATION

California has become a leader in committing to reducing greenhouse gas emissions that contribute to climate change. One approach has been to promote a cap-and-trade system in which those generating CO_2 emissions (polluters) are assigned an amount of greenhouse gases they can

How Woody Plants Sequester Carbon

The process of photosynthesis combines atmospheric carbon dioxide with water, releasing oxygen into the atmosphere and incorporating the carbon atoms into the cells of plants. Soils also capture carbon. Trees, unlike annual plants that die and decompose yearly, are long-lived plants that develop a large biomass, thereby capturing large amounts of carbon over many decades. Thus, a forest ecosystem can capture and retain large volumes of carbon over hundreds of years.

Trees operate both as vehicles for capturing carbon and as carbon reservoirs. Trees sequester carbon in rough proportion to their growth in biomass. An old-growth forest acts as a reservoir, holding large volumes of carbon for hundreds of years. The Intergovernmental Panel on Climate Change (IPCC) states that forest stewardship directed at carbon sequestration could make a significant difference in global carbon sequestration over the near and medium term. Strategies for sequestering carbon in wood products include laminated beams made from smaller trees that are glued together. The small trees may come from thinning projects, a double win. Glue-laminated beams can replace steel beams, which have a higher embedded carbon footprint.

Adapted from Roger Sedjo, 2001. "Forest Carbon Sequestration: Some Issues for Forest Investments." RFF Discussion paper 01-34, Resources for the Future, Washington, DC.

emit (pollute). The amount (the cap) declines each year, so fewer allowances remain as time goes on. If a capped entity, such as an electric utility or natural gas provider, cannot reduce emissions below its cap or if it is very expensive to reduce its own emissions, it can choose to buy allowances from other capped entities. This is the "trade" of *cap-and-trade*. As of spring 2024, 81 percent of the "offset" credits purchased by carbon emitters subject to the cap-and-trade program went toward forest protection projects certified by the California Air Resources Board (CARB), which tracks these offset purchases. While it is hard to predict how effective cap-and-trade programs will be in actually slowing climate change, California and CARB are at the forefront of implementing this "carbon market" approach, and the offset program has generated funds for the protection of forestlands.

One of the biggest ways to remove and store carbon is just to keep trees standing! Especially old-growth, large trees that remove from the atmosphere and store hundreds of tons of carbon each. If trees are cut,

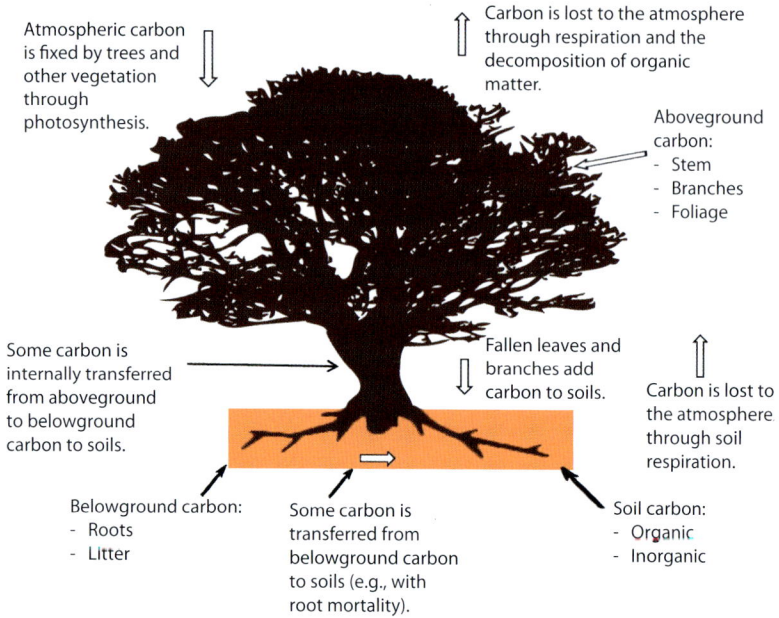

Atmospheric carbon is fixed by trees and other vegetation through photosynthesis.

Carbon is lost to the atmosphere through respiration and the decomposition of organic matter.

Aboveground carbon:
- Stem
- Branches
- Foliage

Some carbon is internally transferred from aboveground to belowground carbon to soils.

Fallen leaves and branches add carbon to soils.

Carbon is lost to the atmosphere through soil respiration.

Belowground carbon:
- Roots
- Litter

Some carbon is transferred from belowground carbon to soils (e.g., with root mortality).

Soil carbon:
- Organic
- Inorganic

Carbon in living systems. Carbon moves among living organisms, the soil, and the atmosphere. Photosynthesis harvests carbon from the atmosphere and stores (sequesters) it in living and dead organisms and in soil. Decomposition releases carbon to the atmosphere. Adapted from the US Environmental Protection Agency.

maximizing the use of the timber and timber by-products is important. There are several types of innovative wood products being made and used in California, including cross-laminated timber and mass plywood panels that may have an impact on the carbon footprint of buildings. San Francisco now has a multistory building that was constructed using cross-laminated and glue-laminated timber and that offers sustainable and attractive office space and light industrial space. There is still a significant amount to comprehend regarding the circumstances and methods through which mass timber production can aid in forest restoration, improve the utilization of small trees, and reduce the use of high-carbon cement and steel in building construction, but the future looks bright.

Other recent changes include an increased interest in forest certification—an assessment of whether individual forests are being managed sustainably—and in new commodities harvested from the forest. Called nontimber forest products, these specialty items include mushrooms, boughs, moss, maple sap, and ferns. There is now greater public input into decision-making, and how management actions affect local people and communities is factored into the decision-making process. There is greater landscape, rather than stand-level, planning, as well as efforts to protect habitats rather than individual wildlife species.

Though the forests and woodlands of California have undergone dramatic changes, they are still the dominant feature in large portions of the state. It's rewarding to continue the long tradition of tending these areas with fire, removing invasive plants, and harvesting the foods they offer, such as acorns, mushrooms, and huckleberries.

RANGELANDS AND LIVESTOCK GRAZING

The most common use of California's rangelands has been livestock grazing. Livestock continue to play a strong cultural role, produce important economic products, and are an important tool in land management. Domestic cattle, sheep, and horses arrived with Spanish colonists in 1769, and by the early 1800s livestock grazing was widespread throughout California. During the Gold Rush, livestock grazing boomed because of an increased demand for meat, and by the late 1800s there were about 3 million cattle and 6 million sheep on California's grasslands.

The US Department of Agriculture has estimated that about 5.2 million head of cattle (and 500,000 sheep) are in the state today, making California the fifth-largest cattle-producing state in the nation. In 2023 the California cattle industry was valued at $4.76 billion from around 11,800 ranches across the state.

Cattle grazing on private land helps preserve 20 million acres of open rangeland landscapes. Equally important, land trusts are actively working to conserve rangelands as open space. Conservation easements can be used to protect these working lands by restricting some or all of the development rights and yet allow ranchers to continue with their important work.

CONSERVATION BIOLOGY

Conservation biology is a relatively new and interdisciplinary science. It has its roots in ecology and wildlife management and emerged from the realization that the Earth is in a biodiversity crisis. Its focus is on maintaining the Earth's biodiversity, including natural ecological and evolutionary processes. To achieve this goal, conservation biologists work at all scales of life, from genes to ecosystems. There is increasing recognition among conservation biologists that social science is integral to

Restoring California's Native Grasses

California's grasslands were once vegetated by native perennial grasses. But during the last 200 years, exotic annual grasses from Europe have largely replaced them, and now only 2 percent of the state's grasslands are vegetated by native perennial grasses. There are about 300 species of native grasses, which began to be displaced when Spaniards settled in California in the 1700s, bringing livestock, Eurasian grasses, and new land management practices. Annual grasses took over from native perennial grasses in the 1800s, largely as a result of overgrazing.

A study of three grass fields 30 miles west of Sacramento, California, was conducted to increase our understanding of how well revegetation with native grasses works. One field contains annual grasses, one is a newly restored field of native grasses, and the third has contained native grasses for 10 years. "We're looking at various methods, such as using fire, applying herbicides, and determining what species grow best in which areas of California," said Stephen M. Griffith with the USDA.

Unlike annual grasses, perennial grasses turn green faster, stay green longer, resprout quickly after fire, and produce more biomass, which can benefit soil development, wildlife, and livestock. This equates to more protein and higher-value forage for both wildlife and livestock. The site that contained annual grasses had significantly fewer tons per acre of aboveground plant biomass accumulation than the two plots containing native grasses. Native grasses integrate better with other California native plants. That diversity of plants attracts wildlife not found in annual grasses. "We've noticed a significant change in biodiversity," said California farmer John Anderson about the native planting experiments. Furthermore, native grasses improve soil and limit erosion.

One constraint is that native-grass seed is very expensive. It may cost $40 a pound while turfgrass seed costs about 50 cents a pound. In the absence of government subsidy of native-grass seeds, perhaps the demonstrated positive effects of native grasslands will influence supply and demand and make native grass more affordable.

Adapted from David Elstein, May 2004. *Agricultural Research*.

providing solutions for the biodiversity crisis. This includes working with Indigenous Peoples, land use planning, landscape architecture, political ecology, and other disciplines. Michael Soulé, who helped found the Society for Conservation Biology, characterized conservation biology as a "crisis discipline," in which tactical decisions must sometimes be made in the face of uncertain knowledge. The discipline has

All You Need to Know to Be a Principled Ecologist

- All species in ecological systems are dependent upon other species for their existence.
- The organisms in ecological systems nearly always act to maximize their individual fitness (reproductive success), not to benefit the population, community, or ecosystem.
- Change is the norm at all levels of organization.
- While each successively larger scale is composed of the units of the next smaller scale, it possesses properties unique to that scale.
- Ecosystems are altered by human manipulations.
- The abundance and distribution of a species will depend on its interaction with other species and with the abiotic environment: soils, wind, temperature, light, water, geologic formations, and natural and human-induced disturbances.

Adapted from Mary Orland, "Principles of Ecology," *Essays on Wildlife Conservation* (Peter Moyle and Douglas Kelt, eds.) at www.marinebio.org.

developed ways to address complex environmental issues with dynamic solutions arising from evidence-based research.

How is it, exactly, that human activity impacts natural communities? Does cutting an oak or patch of oaks diminish the value of the remaining woodland? It depends. Rarely do such acts happen in isolation. It is the overall pattern of change that dictates the level of impact on a plant community or on a population of animals that depend on those plants. Certainly, human activities across all scales, from local harvesting to global climate change, can result in loss of individual species and/or loss of habitat. But it is often the cumulative effect of many small patches of habitat lost that affects the quality of an ecological community.

When a patch of continuous habitat is broken up into many small patches, in a process called fragmentation, the remaining patches may not be able to support all of the species the original patch did. Even when habitat loss is due to natural causes such as hurricanes or earthquakes, smaller habitat patches contain fewer species, as well as fewer specialists (that is, species that depend on specific habitat, foods, or relationships in order to survive). This change in patch size, with a reduction in interior or core habitat and an increase in edge habitat, has implications for the conservation of biodiversity.

Ecological Values and Land Use

The Ecological Society of America (ESA) has prepared a set of basic guidelines that provide a conceptual framework they would like to see applied to land use planning. They recommend that the following checklist of factors should be considered in making land use decisions:

1. Examine the impacts of local decisions in a regional context. If everyone did what you are thinking of doing, would it be sustainable?
2. Plan for events that occur infrequently, such as 20- and 100-year storms!
3. Preserve rare landscape elements, critical habitats, and associated species. First, you will need to know what is rare and common in your area!
4. Avoid land uses that deplete natural resources.
5. Retain large contiguous patches of habitat. Maintain and restore connectivity among patches, and protect and restore rare species and critical habitats.
6. Minimize the introduction and spread of nonnative species—it helps to buy locally grown food and locally made products.
7. Avoid or compensate for effects of development on ecological processes.
8. Advocate for land uses that are compatible with the natural potential of the area and nature-based solutions to increase resilience.

Adapted from V. H. Dale, et al., 2000. "*Ecological* Principles and Guidelines for Managing the Use of Land." Ecological *Applications* 10(3): 639–670.

Size and isolation of patches are two of the most important factors affecting whether species persist in an area after fragmentation. Larger patches support larger numbers of species. In addition, the relative isolation of those patches from one another is important because, ultimately, many species that require large areas to maintain functional populations need to move among the remaining habitat patches to survive, whether many small patches or several large patches of habitat remain. Patch size and location relative to one another, as well as how permeable the landscape is between patches, determine habitat connectivity, a measure of the extent to which plants and animals can move among habitat patches. High levels of habitat connectivity may lessen

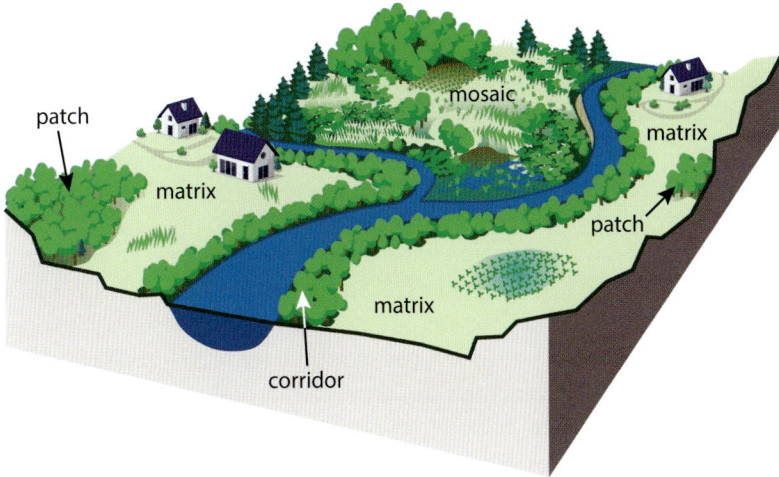

Fragmentation vocabulary: look at this simplified graphic and see if you can explain the words *matrix*, *mosaic*, *patch*, and *corridor*. Adapted from the Federal Interagency Stream Restoration Working Group.

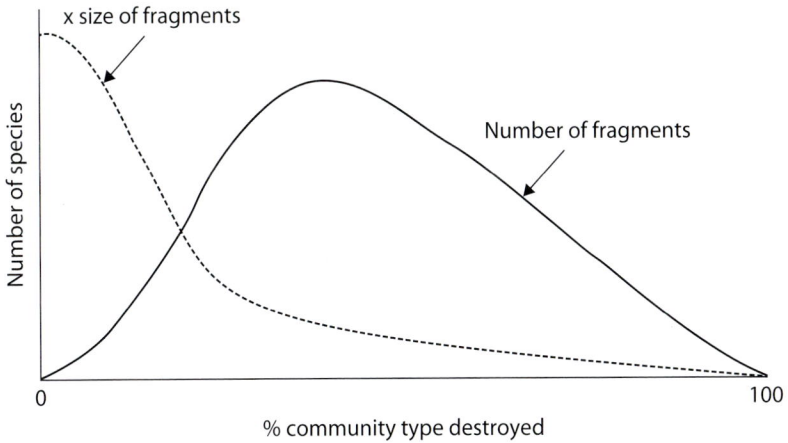

Community fragmentation. The number of species decreases as the number of fragments increases. The number of species also decreases as the size of fragments decreases. From *Corridor Ecology: Linking Landscapes for Biodiversity Conservation and Climate Adaptation*, 2nd edition, by Jodi A. Hilty, Annika T. H. Keeley, William Z. Lidicker, and Adina M. Merenlender. Copyright © 2019. Reproduced by permission of Island Press, Washington, DC.

some impacts of fragmentation by allowing individuals or populations to use more than one habitat patch, thus helping species persist in a region. Valley oak woodland provides an example of how patch size, isolation, and connectivity impact species conservation. Because habitat loss has occurred across a large portion of the valley oaks' historical range, many of the remaining stands of valley oaks are small. This means that an acorn woodpecker wishing to use many trees to collect acorns may be unable to move long distances between unconnected patches, and thus the woodpecker disappears from the isolated patches.

What does this mean for the future of conservation in California? Scientists throughout California and beyond have been looking into this question. Using the principles of conservation biology, scientists have begun to prioritize sites for ecological corridors to protect habitat connectivity. Corridors range in size from road crossings such as the Wallis Annenberg Wildlife Crossing outside Los Angeles—designed to facilitate animal movement over US Highway 101 in Southern California—to landscape linkages from Baja to British Columbia, for example. What are the quality, size, and characteristics of a particular habitat patch? Where is it located in relation to similar habitat patches? Is it at high risk for conversion to urban development or agriculture? Is it a large enough patch of habitat to allow native species to persist? These are some of the questions that we must address to prioritize protection and corridor implementation to protect California's ecosystems and the species they harbor. Open space and habitat connectivity are vital for biodiversity in wild and urban landscapes. As the Urban Greenspaces Institute notes: "In livable cities is preservation of the wild."

Explore!

VISIT A NEARBY NATURAL AREA
Go to the visitor's center and find out what, if any, human activities have influenced the landscape, then look for the "evidence" as you walk the land.

VISIT UNUSUAL FORESTS
We are fortunate to live in a state with some of the world's largest trees and forests. Don't let more time pass without visiting some of them! Good places to start include these:

(continued)

- The Staircase at Jug Handle State Natural Reserve has a series of terraces, each one featuring pygmy trees.
- Redwood National Park is home to some of the tallest coast redwoods (*Sequoia sempervirens*).
- Sequoia and Kings Canyon National Parks are home to the largest trees on Earth, including a giant sequoia (*Sequoiadendron giganteum*) called General Sherman. These parks are also great places to see prescribed fire management.
- Demonstration forests are operated by CAL FIRE throughout the state for research on forest management techniques. There are eight, and they're open to the public.
- US Forest Service Research Natural Areas are often created to showcase interesting or unusual forest types.

ENGAGE IN LOCAL LAND USE PLANNING

Learn about land use, zoning, and development in your area. Who are the stakeholders and what are their positions? What is the history of land use and development? What does your county's general plan say about development of rural parcels? Research zoning, grading, and tree preservation ordinances to see what protections these policies afford woodlands and forestlands. For the politically inclined, attend meetings of your planning commission, city council, or county supervisors.

PLANT TREES AND REDUCE DEFORESTATION

Planting is an excellent way to learn about local trees. Many local, county, state, and federal parks and forests, as well as California ReLeaf, American Forests, and the Arbor Day Foundation, have organized tree-planting events you can join.

PRACTICE FOREST BATHING

Forest bathing was popularized in Japan in the 1980s as a way of communing with nature. Called *shinrin-yoku* ("forest bathing" or "taking in the forest atmosphere"), it offers an antidote to our attachment to our phones by encouraging people to reconnect with and protect forests.

GET CREATIVE

Spend some time with a tree or in a forest, then create a piece of art based on your experience—a song, a story, photographs, poems. In the fall, weave a garland of fallen leaves.

6

Animals

Observing animals can stir a range of emotions within us. From the fondness we feel for a bush bunny to the excitement of spotting an eagle soaring above us, each experience brings its own mix of feelings. Observing animals can immerse us in a state of flow, where time seems to slip away, leaving us with a lasting sense of joy. This chapter will lead you through the world of animals, from those living belowground to those flying above us, and how they are related and interrelated in complex communities of life. While we may find it easier to relate to other vertebrates, it's important not to overlook the significance of invertebrates, because they play a crucial role in ecosystems and in our own survival. Tuning into the little things is something naturalists do. Insects can provide a never-ending source of wonder but also need our conservation attention as their numbers are dwindling in the face of global change. Keep reading to get to reptiles and amphibians, which are sometimes hard to find but well worth the search. Mammals such as bears and sea otters offer genuine thrills, but the more elusive bats, nocturnal flying mammals, are also really cool. We hope this chapter will help you get into the animal flow.

Animals as sentient beings with agency and intelligence are deeply embedded in Indigenous culture, where animals frequently play central roles in traditional stories and narratives that explain existence and ecosystem processes and guide ways of living. Imagine the coyote as the trickster, and picture the eagle, high in the sky with its gaze piercing the veil between worlds. The animals make up a compass that teaches us how to be and how not to be. When boundaries between the human and animal realms are blurred and their narratives intertwined, the intrinsic relationship between humans, animals, and the broader ecosystem is affirmed from generation to generation. This can foster a deep sense of kinship and responsibility toward all living beings and underpin important rules that strengthen mutual dependence, respect, and reciprocity.

The Bat Brothers Banish Warm Wind

Turtle blinked her shiny eyes, then said, "Warm Wind was once married to Mole. Mole is the father of the seven sisters. But Mole proved to be a rather untrustworthy and disloyal husband. So, after a while, Warm Wind left Mole and raised the seven daughters by herself. Now she is getting older and the daughters take good care of her. Indeed, the daughters are good women. Without them Warm Wind would not be able to make her trip back and forth to the large eastern valley, where she goes every year to visit her brothers and sisters. Her father is the sun, which is why when she is below the Mountain we feel her warmth."

Excerpt from "The Bat Brothers Banish Warm Wind" in *How a Mountain Was Made* by Greg Sarris, Heyday Press

ANIMAL ECOLOGY

Animals can be found in different habitats, depending on natural physical processes and species interactions that favor some, exclude others, and change constantly. Predation and competition for resources like food and shelter are good examples of species interactions that structure animal communities. An example of predator-prey interactions is mountain lions hunting and eating mule deer. Deer make up a big part of a lion's diet.

The presence of mountain lions helps to limit deer populations, which in turn affects vegetation in diverse habitats, such as forests, deserts, chaparral, and grasslands. By reducing deer numbers, lions reduce the browsing pressure on many plant species, including oak and bay seedlings. Hence, mountain lions help maintain the health of California's plant communities. Moreover, fear of mountain lions influences

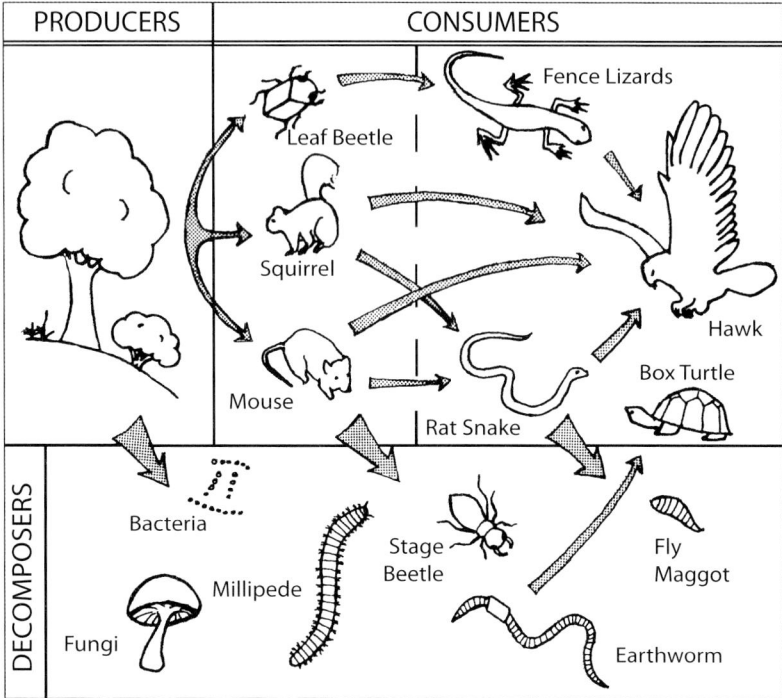

A food web. Producers (photosynthetic plants) are the energy basis of all food chains. From *Conservation Planning: Preserving Ecosystems in Open Space Networks,* University of California Agriculture and Natural Resources publication 3370.

deer behavior, changing their activity patterns and habitat selection, which can further impact vegetation dynamics and biodiversity. This predator-prey relationship illustrates the intricate balance in California ecosystems and highlights the important role that apex predators play in regulating prey populations and shaping ecosystems. Predators and their prey are part of the food web—an important concept for understanding the energy relationships among organisms. A food web graphically describes the many interconnected energy relationships among species in a community including plants, herbivores, carnivores, and microbes. The arrows in the graphic represent the direction of energy transfer, that is, who is eaten by whom. Look at the food web diagram to see examples of producers, consumers, and decomposers.

Which is an herbivore? Is there an omnivore? Notice how the mouse, for example, is a food source for several types of animals. Although a food web is always oversimplified, it attempts to convey the complex interactions that take place in nature.

Animals can be categorized in a variety of other ways in addition to energy relationships: by their lifestyles and activity patterns, their ecological functions, their reproductive strategies, and their unique and wonderful adaptations. For example, all of the animals on Earth can be categorized according to when they are active. Most animals have a typical time of day for being active and a habitual time to rest or hide. Those that graze or hunt during the day are called diurnal, like cows and swallows. Those that are active at night are called nocturnal, with examples being raccoons and spotted owls. There is another group that is not as clear-cut as the above—those that are most active at dawn and dusk but may be somewhat active into the day or the night. They are called crepuscular. Examples might be coyotes, nighthawks, and hawk moths. On a longer scale, some animals are active year-round (people, elk, sharks), while others take a metabolic vacation during the unfavorable season. This time off can look very different for different kinds of animals.

Many people will be familiar with the idea of hibernation, where an animal becomes inactive; its body temperature, respiration rate, and heart rate are reduced, and its metabolism is greatly slowed down, usually in relation to cold temperatures and short day length. Hibernation is an example of an endothermic (warm-blooded) mammal taking an extended break. Other examples of metabolic change that may be less familiar are birds and bats going into a state of torpor for half the day (bats during the day and some birds overnight). In the torpid state, birds or bats have reduced core temperatures and lower heart and respiration rates, but this state lasts only about 12 hours. This ability apparently helps these small endotherms to conserve energy. Essentially, while they sleep, they save the cost of heating, and then they have that energy for flight activity the following day. The Anna's hummingbird you might see in your garden may drop its body temperature overnight by about 50 to 60 degrees! Examples of hibernation in ectothermic (cold-blooded) animals are far more dramatic, such as mountain yellow-legged frogs spending the winter below the ice in a frozen-over lake in the Sierra, and fish that are able to come out swimming after the ice melts. Many ectothermic vertebrates spend extended periods being inactive, either during the cold, short days of winter (hibernation) or during the hot, dry days of summer (estivation). Red-legged frogs, slender salamanders, and many others retreat to deep underground cavities to avoid desiccation during the 6-month California dry season.

Why We Need Sea Otters and Other Top Carnivores

Sea otter (*Enhydra lutris*) near Morro Bay, California. Photo courtesy of Michael L. Baird, flickr.bairdphotos.com.

Sea otters have not had an easy time of it along the North Pacific coast. By 1900, populations were decimated by the fur trade. Their absence in coastal waters has contributed to unraveling ecosystems. Sea otters are top predators and have a critical effect on trophic levels below them. They eat shellfish, including sea urchins. Sea urchins feed on kelp. Kelp makes up the forest of the sea, providing habitat and food for myriad other life-forms. When sea otters are missing from the coastal reefs, unchecked urchin numbers feed on the kelp until there isn't any more. Lack of kelp eventually means that small fish and shellfish go too. The productivity of the marine system without its top predator is vastly curtailed.

(continued)

In the lowlands of California, estivation is more common and important, whereas in the higher-elevation areas of the Sierra Nevada, Cascades, and Southern California mountains, hibernation is more important. Another approach to avoiding the unfavorable season is to have a life cycle oriented around the seasons and to pass the unfavorable season as a hard-shelled egg, a desiccation-resistant pupa, or an inactive grub. Insects are specialized for all of these techniques, and spiders, centipedes, and millipedes exhibit some of them.

Why We Need Sea Otters and Other Top Carnivores *(continued)*

Efforts to recover sea otters in coastal California have had many beneficial impacts. Agricultural runoff brings a nitrogen overload to Monterey Bay's Elkhorn Slough, and this accelerates algae growth. Algae smother eelgrass, preventing it from photosynthesizing. Like kelp, eelgrass is a major structural element of the ecosystem and a prime habitat for sea otters. But with the sea otters gone, sea slugs and other algae-feeding animals could not keep the algae in check because crabs were eating the sea slugs. Once the sea otters were reintroduced, they stepped up to the proverbial plate and began dining on the crabs. Fewer crabs led to more sea slugs led to less algae. The presence of sea otters helped restore eelgrass that is home and hearth to many more species.

Coastal ecosystems can sequester 10 times more carbon than tropical forests and hold that carbon much longer. These "blue carbon" ecosystems are under assault, and globally we lose almost 2 percent of seagrass meadows every year. Accelerating our efforts to recover sea otter populations could help restore coastal ecosystems and advance resilience to climate change.

Mary Ellen Hannibal, 2024.

It is important to remember that life on Earth arose in the ocean, and only much later were various forms of life able to colonize the great land masses. Thus, all terrestrial life on Earth, which we consider the norm, is in fact highly specialized and derived from aquatic ancestors. Much of the selection pressure that constrains the forms of animals on land today is related to the difficulty of existing in a terrestrial habitat and coping with water loss. Insects are characterized by having a hard outer shell (an exoskeleton) made of chitin. It can be argued that the most important evolutionary function of the insect exoskeleton is to retain moisture. The skin of humans is far less efficient than an insect exoskeleton at impeding water loss. Nonetheless, one of the main functions of our skin is to keep us from drying out.

Animals are terrific architects of the environment. Grasslands and meadows remain open grasslands mostly because bison, elk, and gophers keep them that way. Without the grazing and trampling effect of large mammals, and the soil-rotating effect of small burrowing mammals, most meadows would soon become shrublands and woodlands.

Ladybird beetle winter congregation. Ladybird beetles migrate to winter hibernation masses, feed on aphids, and are chemically protected. Photo by H. Vannoy Davis © California Academy of Sciences.

Beavers are a startling example of animals that manipulate habitat to suit their needs and end up changing the world. In addition to felling trees and opening up grazing habitat to sunlight, beavers dam creeks and small rivers, turning flowing-water habitat into still-water habitat. Slowing water flow and forming pools can have benefits for the many species dependent on water, including us, as the heat of summer takes hold. The water temperature and water chemistry, the aquatic plants, the fish, and every other aspect of the habitat are altered. Whole suites of organisms are excluded or invited when a beaver dams a river. The beaver is a good example of how important animals are for the destruction and recycling of plants (decomposition). Without the beetles that eat dead trees, for example, the entire world would be a giant pick-up-sticks-like maze of fallen, dead, undecomposed trees.

Perhaps the most enthralling aspect of watching wildlife is the diversity of behaviors one sees. Picture a black-tailed jackrabbit darting from shrub to shrub, evading predators such as coyotes, rattlesnakes, red-tailed hawks, and eagles. And now compare that with the strategy of the Gila monster, which relies on its ability to remain perfectly still, taking advantage of its camouflage coloration to blend in with the sub-

Bighorn sheep in Southern California. Photo by Gerald and Buff Corsi.

strate to protect itself from predation. Each animal will display behaviors of some kind, from sitting quietly to avoid detection to making dramatic noises or movements that serve a suite of evolutionary or life history needs. Behavior is a large and fascinating topic and provides the action many people enjoy viewing in film and photographs. It takes time to observe animal behavior in the wild, but the rewards are legion.

EVOLUTIONARY GROUPS

One way to think about animals is by examining evolutionary groups and exploring their phylogeny, which shows relationships based on shared origins. Although scientists today are far from unanimous in using the six-kingdom scheme shown in the table to describe the diversity of life on Earth, it is useful. The six kingdoms are Bacteria, Archaebacteria, Protista (protists), Fungi, Plantae (plants), and Animalia (animals). The organisms in the Bacteria and Archaebacteria kingdoms are prokaryotes, which have simple cellular structures that lack organelles or a nucleus and usually reproduce asexually. A few prokaryotes may be slightly familiar thanks to the diseases they cause; for example, *Streptococcus pyogenes* is the source of strep throat. The organisms in the other four kingdoms are eukaryotes, which have organelles and a

TABLE 6.1   KINGDOMS OF LIVING THINGS IN THE LINNAEAN CLASSIFICATION SYSTEM

Kingdom	Structural Organization	Methods of Nutrition	Types of Organisms	Total Species (estimated)
Bacteria	Single prokaryotic cell (organelles like the nucleus not enclosed by a membrane)	Absorb or make food	Bacteria, cyanobacteria (also called blue-green algae), and spirochetes	Millions to billions
Archaebacteria	Single prokaryotic cell (organelles not enclosed by a membrane and different cell wall structure than in Bacteria)	Absorb or make food	Found in extreme environments	Millions to billions
Protista	Single eukaryotic cell (nucleus enclosed by a membrane)	Absorb, ingest, and/or make food	Protozoans and algae	200,000+
Fungi	Multicellular, filamentous forms with specialized eukaryotic cells	Absorb food	Fungi, molds, mushrooms, yeasts, and mildews	2–4 million
Plantae	Multicellular forms with specialized eukaryotic cells without their own means of locomotion	Produce food	Mosses, ferns, conifers, and flowering plants	435,000+
Animalia	Multicellular forms with specialized eukaryotic cells with their own means of locomotion	Ingest food	Invertebrates, fish, amphibians, reptiles, birds, and mammals	1.5 million+

nuclear envelope in their cellular structures. The eukaryotes can be divided into the three easily recognized kingdoms composed of fungi, plants, and animals, plus the mostly unicellular protists.

The division of kingdoms is clearly artificial in some ways, stressing large multicellular animals and plants. At the root of the tree of life there is tremendous diversity among unicellular organisms that the six-kingdom scheme does not recognize. This system divides animals into two major groups: invertebrates and vertebrates. Invertebrates include worms, clams, spiders, scorpions, millipedes, centipedes, insects, and many other animals without bones. The invertebrates are by far the larger animal group, in terms of total number of species and in terms of global biomass, yet generally receive less attention than the vertebrates.

Vertebrates are the group with fish, birds, frogs, mammals, toads, turtles, lizards, snakes, crocodilians, salamanders, and dinosaurs. The principal feature distinguishing these two great animal groups is the presence or absence of a bony or cartilaginous backbone: vertebrates have one, invertebrates do not. Invertebrates generally utilize an outer shell, or exoskeleton, to perform the functions vertebrates rely on internal bony or cartilaginous structures for.

INVERTEBRATES

The invertebrate world is as strange as any Hollywood horror movie, if not more so. Within the great group of invertebrates, evolution appears to proceed along a simple line: the differentiation of a segmented body plan. Imagine the most ancient ancestor of spiders, crabs, millipedes, and beetles looking like a simple, undifferentiated worm with legs. The body is externally supported, with internal organs like lungs, circulatory system, and muscles being attached to the exoskeleton. It is bilaterally symmetrical, with each segment of the body bearing one pair of appendages. Each of the modern invertebrate forms has been derived by taking the original body plan and reducing, increasing, or merging the number of segments and modifying the appendages for specific functions.

People sometimes refer to all "creepy-crawlies" as "insects," when actually insects are only one of the types of invertebrates. Invertebrates are a highly diverse lineage and include spiders, crabs, scorpions, and many other invertebrates that are not insects. Here we describe some common groups of California invertebrates, identify special reasons for noticing them, describe a few of the distinguishing features of these

Click beetle. Head and thorax are merged. Antennae are beaded, with triangular beads, the first much larger. Eyes protrude from the side of the head. Six segmented legs. Elytra cover the abdomen, hiding the folded wings below them. © Joyce Gross, http://joycegross.com.

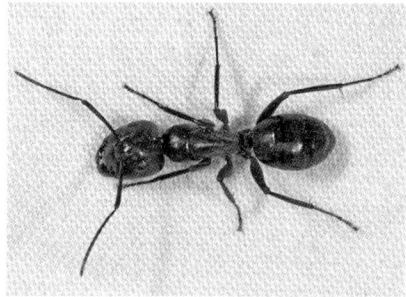

Ant. Head (left), thorax and abdomen (right) visible. Six legs attached to the thorax, with six distinct segments, final segment with paired claws. Head wider than the thorax. Antennae attached to the head, basal segment far longer than the others. © Joyce Gross, http://joycegross.com.

groups, and highlight a few of the most common organisms and the patterns they exemplify.

Centipedes are perhaps the invertebrate most superficially like an ancestral arthropod, and they are good organisms to use in imagining the ancestral invertebrate body plan. They are strongly segmented, with all segments except those in the head region bearing only one set of jointed legs. Thus, they exhibit little evidence of merging or modification of segments. Centipedes are flattened, which facilitates their habit of living under rocks, logs, and bark and moving through tight spaces. Unlike insects, centipedes have no waxy coating on the exoskeleton; thus they are highly susceptible to drying out (desiccation).

For this reason, they are strictly nocturnal, prefer to live in damp microhabitats, or are active only during appropriate climatic conditions. In other words, they are found in moist forests as well as under rocks and garbage cans in the wet season in most of California, but they go deeper underground and become inactive in the dry season. The centipede head bears a pair of appendages highly modified for chemical and tactile sensory function (antennae) but has no eyes or only simple eyes (ocelli) that tell light from dark but do not resolve an image. So, these highly active, fast-moving predators are blind! They hunt by smell

Millipede. Two beaded antennae on head (right). Each body segment (ring) has two pairs of legs. Millipedes are chemically defended. © Joyce Gross, http://joycegross.com.

and feel. They take 1 to 3 years to grow from an egg to an adult, then live 3 to 6 additional years. This is a long-lived invertebrate. Although adult centipedes may have 15 to 177 pairs of legs and typically have an odd total number of legs, they are born with only 3 to 7 pairs. With each molt they add pairs of legs until they reach adulthood.

Millipedes are another classic, segmented arthropod, but this time a detritivore-herbivore with two pairs of legs per segment. Thus, this animal exhibits a simple variation on the segmented-worm-with-legs theme: merging without reduction. Each modern segment of a millipede's body represents two ancestral segments that have become one, with the appendages (legs) from both ancestral segments retained. Millipedes are cylindrical animals, contrasting sharply with the flattened centipedes. Millipedes are thought to be among the first animals to have colonized terrestrial habitats during the Silurian period (443–419 million years ago, or mya). The oldest known fossil of a land animal, *Pneumodesmus newmani*, was a tiny millipede. There are around 8,000 to 12,000 species of millipedes worldwide.

Crustaceans are a distinctive group of largely aquatic invertebrates that include well-known macroscopic forms like crabs, lobsters, shrimp, crawdads (also called crayfish), and sow bugs (also called pill bugs). Crustaceans also include ecologically important microscopic forms that inhabit fresh water, especially ponds, and form an important part of aquatic food chains (fairy shrimp, water fleas, copepods, and ostracods). One of the distinctions of the crustaceans, being largely aquatic, is that most of them breathe by means of gills.

Spiders, members of the class of arthropods called Arachnida, make up one of the larger, more successful groups of arthropods, with approximately 45,000 to 50,000 described species globally. They are prominent in most habitats and easily observed, and for these reasons they can be very

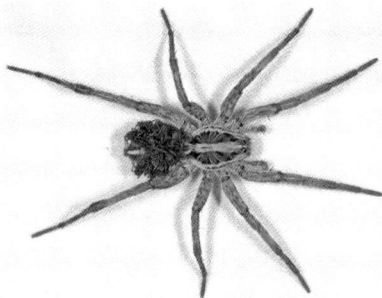

Wolf spiders have eight legs, eight eyes, and a pair of palps (short, leglike appendages). © Joyce Gross, http://joycegross.com.

attractive to naturalists. Spiders have two body segments, the cephalotho- rax and the abdomen. The cephalothorax carries the four pairs of legs, the eight eyes (!), and the chelicerae (fangs). Under the abdomen, toward the rear, are the (usually six) spinnerets, through which silk is emitted. Spiders have a relatively simple life history. They hatch from eggs as tiny versions of the adult. They then shed their exoskeletons to grow larger.

Virtually all spiders have venom glands, which they use to subdue prey or for defense. In most spiders the venom is delivered via hollow fangs, each with a hole near the tip. Most spiders are too small to suc- cessfully inject venom through human skin, and most of the spiders whose fangs are large enough to penetrate our skin have innocuous venom. There are two spiders native to North America that are danger- ously poisonous, the black widow and the brown recluse, but only the black widow is found in California. The black widow is typically shiny black in color with a characteristic red hourglass-shaped marking on the underside of its abdomen. Keep an eye out for them in dark, secluded areas. They are especially likely to be in places where you cannot see them, like under a log. But don't panic; they generally remain hidden and motionless in their webs or hiding spots unless disturbed, relying on their venomous bite to immobilize prey rather than actively chasing it down.

Insects are the most successful group of organisms on Earth, as meas- ured in terms of number of species or biomass. Estimates of the number of species of insects in the world vary wildly, with over 1 million described and the estimated total number ranging between 10 and 30 million. The buzz of a honeybee, the sting of a yellow jacket, the beauty of butterflies, the buzz of cicadas are all familiar parts of daily life.

Insects have such different approaches to life than mammals that it is mind-boggling to contemplate them. The insect body plan is an exoskeleton divided into three segments: the head, thorax (chest), and abdomen (belly). On the head are the sensory and feeding structures, the modified legs that have become antennae, and mouthparts. The thorax carries the three pairs of legs and usually one or two pairs of wings. The abdomen is appendage-less but carries the reproductive parts, which function through the tip (distal end) of the abdomen.

Each species has a lifestyle, and these lifestyles can be grouped into two broad categories: complete and incomplete metamorphosis. Complete metamorphosis is the classic insect life cycle exemplified by butterflies: egg to grub (larva), grub to pupa (chrysalis, or cocoon), pupa to adult. This life cycle (termed holometabolous) is shared by butterflies, moths, beetles, flies, caddis flies, fleas, and wasps. Adult insects do not grow. This fact bears repeating, as it is crucial to an understanding of insect life cycle adaptations: adult insects do not grow, nor do they ever shed their shells again. If one loses a leg, it is gone, never to be regrown. This is unlike spiders, which continue to shed their shells throughout their lives and may regrow lost limbs.

Incomplete metamorphosis is also termed gradual development. In insects exhibiting incomplete metamorphosis, the form that hatches from the egg looks like a small, wingless version of the adult. It then sheds its shell several times, growing larger with each shed (each molt). On the final shed, new structures are added, including wings and reproductive organs. Incomplete metamorphosis is practiced by grasshoppers, aphids, stinkbugs, roaches, and silverfish. There is one major variation on the theme of incomplete metamorphosis, that of some aquatic insects like dragonflies, damselflies, and mayflies (hemimetabolous insects). Hemimetabolous insects grow through a series of molts, but the juveniles look quite unlike the adult form, being wingless and adapted to an underwater life very different from that of the adults. Hemimetabolous insects pass through a "transforming stage." In this stage they are nonfeeding, relying on stored resources to produce the wings and reproductive organs necessary for their next life stage. Reliance on stored resources to power transformation is the same as in insects with complete metamorphosis.

Insects have achieved remarkable diversity and ecological significance through two key adaptations: complex life cycles and the ability to fly. Flight allows insects to colonize distant lands, cross water barriers, pass over mountains, and skip over deserts to find suitable habitat beyond. Flight also gives the advantage of transportation to safe hiding, nesting, and overwintering sites, and escaping enemies. Many insects have incorporated flight into their mating and hunting strategies, and

Native Bees Provide Resilience and Need Protection

Andrena chalybaea, a California native bee, is thought to be monolectic, meaning it generally collects pollen from only one plant species, in this case *Camissonia ovata*. Photo by Rollin Coville.

California's diverse ecosystems harbor an estimated 1,600 species of native bees, many adapted to specific ecological niches. While people often associate pollination with nonnative honeybees (*Apis mellifera*), native bee species not only are essential for pollinating native plants, but also serve to pollinate crops and ornamental plants across the state. Bumblebees and other native bees are declining because of exposure to pesticides. Habitat loss, disease, climate change, and competition from nonnative species also contribute to native bee declines. To help native bees, we can remove turf and plant native plants in our gardens and parks and in and around crop plantings, providing food and some bare ground that is lightly mulched for nesting habitat for these essential pollinators. By choosing organic produce and promoting sustainable farming practices, we can help protect bees and contribute to a healthier environment for the entire community of life. And through education and advocacy, each of us can play a part in ensuring the well-being of California's native bees, and enjoy delicious food!

these early fliers evolved around 400 mya, some 250 million years ahead of birds, and about 350 million years before bats took to the air.

Complex life cycles, where the juvenile may occupy a very different niche than the adult does, are another key to insect success. Consider for a moment a dragonfly hovering above a pond. The adult is winged and lives on the air, hunting, mating, and escaping predators by flight. The juvenile stage is a wingless, wormlike animal that lives in the mud on the bottom of the pond, ambushing passing bugs for its sustenance! Another example is wood-boring beetles, whose soft-bodied larvae bore through and around logs, eating cellulose for dinner, excavating a tunnel to live in, and avoiding desiccation by hiding in the cool, dark, moist tunnel. Meanwhile, the adults fly from place to place in the open air, seeking mates and laying eggs. In some insects, success is achieved by separating the foraging strategies of the adult and the juvenile: mom eats spinach, kids eat iceberg lettuce. In others the entire responsibility for eating is delegated to the juvenile form, while the adults concentrate on reproduction. Adult mayflies, for instance, live only to mate; they do not eat. Butterflies and moths are similar; the adults are simply mating machines. The larvae eat to gain weight; the adults sip only water or nectar to keep from drying out or to procure a few calories.

Certified California Naturalists will come to know the California sister (*Adelpha californica*) as the butterfly on their program pin. This striking species, native to California, is distinguished by its dark-brown wings accented with vibrant orange patches and white spots. This butterfly is commonly found in woodlands because oaks are host plants for the spiny small green caterpillars. The adults emerge in the spring and again in the fall and are fast fliers but can sometimes be found staying still along a stream bank.

Insects play an indispensable role in sustaining human societies. They serve as pollinators for crops, enabling the production of fruits, vegetables, and other vital foods. Additionally, insects contribute to ecosystem health by decomposing organic matter, recycling nutrients, and controlling pest populations. Despite their significance, recent years have witnessed a concerning decline in insect populations worldwide. For instance, the Mojave poppy bee (*Perdita meconis*) survives in one tiny corner of its former range. Habitat destruction, pesticide use, climate change, and pollution have led to precipitous declines in insect abundance, raising alarm about potential cascading effects on ecosystems and human well-being. As stewards of the Earth, we must recognize and address the plight of insects to safeguard the delicate balance of our planet's ecosystems.

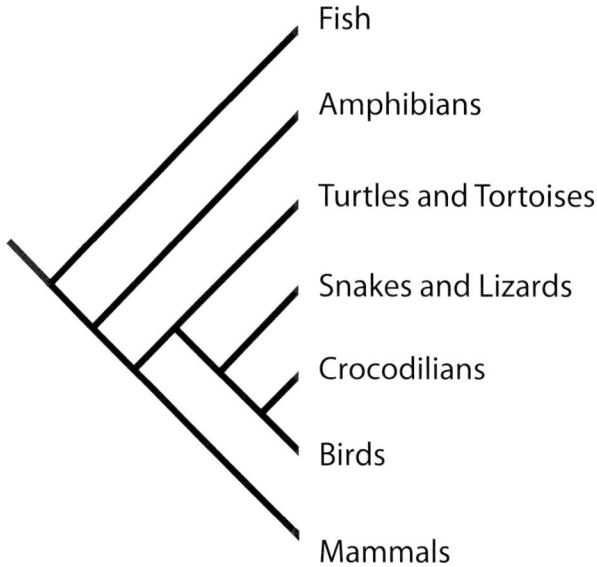

Vertebrate evolution. The branching diagram (cladogram) shows degrees of relationship and ancestry. Groups that branch off together share more traits in common and are inferred to be more closely related.

VERTEBRATES

Many of us have some idea of which mammals are living in our vicinity, but can you imagine which reptiles live within a mile of your home? How many kinds of frogs have you seen or heard? Do you know what fish swim through your town each year? Mammals, reptiles, amphibians, and fish share a common characteristic: they have spinal cords protected in a bony tube. The novel adaptations shared by all vertebrates, especially the internal, bony skeletons used to support organs and muscles, separate this group of animals from the invertebrates. Pause for a moment now to consider whether you can construct a family tree for vertebrates based on what you already know about them. Think about what is most closely related to what, and which evolved from which? Here are the players: snakes, fish, mammals, crocodilians, salamanders, birds, turtles, frogs, and lizards. After thinking about it for a while, take a look at the vertebrate evolution illustration (cladogram), a family tree revealing the common ancestry of all vertebrates.

The first major branch on this tree is fish, which are ancestral to all four-legged creatures (tetrapods). The first tetrapods to distinguish themselves were the amphibians: the frogs, salamanders, and caecilians. Fish and amphibians are distinguished from all further tetrapods by laying nonamniotic (single-layer) eggs. All other tetrapods have amniotic eggs (eggs with multiple layers). Amniotic eggs were a huge advance in the colonization of land in that they allowed amniotes to produce eggs that survived in more and more extreme (dry) environments, as exemplified by the reptiles and birds.

The bird and crocodilian groups are separated from the snake and lizard group by single versus double penises. Yes, lizards and snakes have paired penises! On the crocodilian and bird branches, crocodilians differ from birds (and dinosaurs) in being ectothermic (relying on external surroundings to regulate body temperature) and in having a bony hard palate. Birds separate from dinosaurs, their immediate ancestors (not shown), by the presence of a unique throat modification (the syrinx) that facilitates song, the absence of a jawbone, and lack of teeth. Birds are best thought of as the living descendants of dinosaurs. Birds are, essentially, dinosaurs without teeth! Snakes and lizards are closely related and roamed the Earth long before dinosaurs. In fact, snakes ate dinosaur eggs just the way they enjoy lizard and bird eggs today. The most striking adaptation of snakes, loss of legs, has arisen multiple times. Lizards have also evolved in slightly different guises multiple times.

The last group branching off gives rise to the mammals, with their unique suite of adaptations, notably hair and the production of milk in mammary glands (lactation) to nourish offspring.

Fish

Fish are one of the most ancient life-forms still extant on Earth. Fish first evolved in the oceans and later colonized rivers, lakes, and other aquatic habitats on land. Most animals use oxygen to power their activities, and the ways animals obtain oxygen are critical to their survival. Fish extract oxygen from water through the use of gills. The one-way passage of water over their gills can be continuous and can be powered by swimming (in sharks) or by "pumping" (in most bony fish).

Fish reproduce in a way that may be surprising. In most fish, the female lays eggs that are unfertilized. After the eggs have left the female's body, the male squirts sperm on the eggs. The female and male may never touch in the process of reproduction. Some fish do exhibit internal

Salmon as an Ecocultural Species

Culturally important species represent the intersection of ecological significance and cultural reverence by Indigenous communities and are an important focus for conservation. Salmon hold immense cultural significance for many California Tribes. Their migrations, navigating from rivers to the ocean and back, are celebrated with Indigenous rituals. Salmon provide critical food for the health and connection to place of Tribal communities. Salmon are viewed as sacred beings, essential to the balance and harmony of the natural world, and are part of ceremonies, dances, and rituals. The Yurok use dip nets, seine nets, basket traps, weirs, and spears to take salmon on the Klamath River. They use all parts of the salmon, and they smoke, dry, and ferment fish for storage.

The Utom, or Phantom, River is how Chumash people, with homelands along the central and southern coast, know the Santa Clara River because the flow can come and go unpredictably like a phantom, due to the complex underlying geology and irregular rainfall. The watershed hosts some of the region's most intact and important natural and cultural areas, including Chumash village sites, trading routes, and traditionally tended gathering sites. The Utom River stretches for about 116 miles, originating in the Angeles National Forest atop the San Gabriel Mountains and finally meeting the Pacific Ocean between Oxnard and Ventura. Along its course, it holds traces of Chumash village sites, ancient trading routes, and cherished gathering spots, each reflecting the area's deep-rooted history.

Today Wishtoyo Chumash Foundation honors traditional Chumash knowledge, values, and lifeways, and it provides opportunities for youth to not only feel their connection to the land and water but to engage in building a relationship with the natural world. This effort includes helping to revive endangered cultural keystone species like southern California steelhead, California condor, California red-legged frog, and arroyo toad, safeguarding both the environment and the cultural heritage intertwined with it.

fertilization, especially sculpins, sharks, skates, and rays. A few marine fish produce leathery egg cases that sometimes wash up on a beach, and a few sharks give birth to live young feeding on a placenta-like structure.

Evolutionarily, there are three broad groups of fish in California and its nearshore waters: jawless fish (hagfish and lampreys), cartilaginous fish (sharks, rays, and chimaeras), and bony fish. The most "primitive" surviving vertebrates on Earth are the hagfish and lampreys. These cylindrical, eellike fish are members of the jawless group and are further characterized by having pore-like gill openings along the sides, lacking scales, and lacking pectoral or pelvic fins. Hagfish are rarely seen, bottom-dwelling, deep-water marine forms. Lampreys are born in fresh water, but most species are anadromous; they migrate to the ocean, where they spend part of their lives feeding and growing, then return to rivers to breed. Most species are parasitic or predatory, attaching to the sides of fish or marine mammals and sucking blood or ripping flesh from them.

At one time, the Klamath River Basin in Northern California and southern Oregon harbored the greatest lamprey species diversity on Earth. These fish rich in omega-3 fatty acids provide an important cultural food for Indigenous Peoples, who have a collective memory of lamprey migrations so vast that they left the river's surface covered in a brown film. Today, Yurok people call the lamprey ecotype that mature in the ocean *tewol* ("ocean"), and those that mature in the river are *ke'ween* ("lamprey eel"). Dam construction and resource exploitation have led to a precipitous decline in lamprey population types. Efforts to restore lamprey populations have not received as much attention as projects targeted at salmon recovery, because there is not a commercial lamprey fishery. However, with leadership from the Yurok Tribe, lampreys are receiving increased conservation attention, including research on their biology and migration patterns, improving fish passage through dam modifications, sustainable harvesting practices, and fostering community involvement in conservation efforts.

Sharks and rays are cartilaginous fish and don't have bones. The skeleton to which their muscles are attached is made of cartilage. They are probably familiar to most of us, at least conceptually. Who hasn't seen movie or television footage of a shark? Some of our most beloved films star sharks, if not always in a way that accurately portrays their natural history. Sharks and rays are further differentiated from bony fish by having multiple external gill openings (sharks, five to seven; bony fish, one). Shark skin also differs from that of bony fish. Most bony fish have overlapping scales covering their bodies, while sharks have toothlike denticles that give them a characteristic rough texture.

Rays differ from sharks in being strongly flattened, helpful for bottom dwelling, and in having their pectoral fins attached to their heads to form a characteristic "disk" shape. Rays don't take in water through their mouths for respiration; rather, they take it in through an opening (spiracle) on the upper surface of the "face," which facilitates life on a muddy or sandy bottom. Sharks and rays are generally marine dwellers, though a few enter fresh water. Rays have evolved novel adaptations, including a barbed sting (stingrays) and the ability to produce electric current (electric rays) to defend themselves.

The third group in the cartilaginous fish category is the chimaeras. These are ancient, deep-water marine forms rarely seen. They have one external gill opening; smooth, scaleless, slippery skin; and a large, rabbit-like head. Interestingly, these fish have special electroreceptors, which look like dots on their snouts, to help them sense prey. The common names for these fish mostly focus on the snout—for example, long-nosed, short-nosed, and plough-nosed chimaeras.

Most familiar forms of fish, such as trout, salmon, and aquarium fish, fall into the third evolutionary category, the bony fish. Bony fish occur in virtually all waters of the world, from high altitude (15,840 feet) to deep-sea trenches 36,089 feet below sea level. This group includes the brilliantly colored coral reef fish, the strange monsters of the deep ocean, and most of the fish commonly eaten or cast after for sport. One of the adaptations that evolved in many bony fish is the ability to exploit both fresh and salt water during different portions of their life cycle. This ability to be anadromous is illustrated by salmon, whose eggs hatch under gravel in streams. The fry (small fish) move downstream to the ocean, where they feed and grow for 1 to 5 years. The fully grown adult salmon then swim back up the river in which they were born to spawn (mate and lay eggs). The adults die after spawning, and their carcasses fertilize the river and the adjacent forests. In other words, salmon are harvesting marine resources and delivering them to fertilize freshwater and terrestrial ecosystems!

Coho salmon live in rivers from Monterey, in central California, to Point Hope, Alaska. Many coho salmon runs are imperiled and hence are receiving a good deal of attention focused on their recovery. In contrast, the Lost River sucker and the shortnose sucker are endemic to the headwaters of the Klamath River Basin. These two suckers grow to 3 feet long, weigh 50 pounds, and live to be 40 years old. During the latter half of the twentieth century, both of these species experienced such significant declines that they were listed as endangered under the federal Endangered Species Act in 1988, despite previously being abundant enough to sustain commercial fisheries. Because of their restricted distribution, these species are more susceptible to extinction than the coho. They suffer from conversion of the wetlands around Klamath Lake into agricultural fields, which the juveniles cannot use in place of the lake's clean water and rearing habitat. Perhaps because they are no longer commercially viable, their plight remains relatively unknown. However,

Desert Pupfish

In the Mojave Desert live strange tiny fish unlike most fish you are used to. Desert pupfish live in isolated patches of water habitat, tiny springs, seeps, and caverns in the desert. The water in these isolated pools can be really hot (110 degrees!) and very salty, really challenging conditions for fish. Even worse, conditions can change, daily or seasonally, and the 2-inch-long fish have to put up with it! Water temperature in some pupfish holes may go up and down 35 degrees in a day. The ancestor of our modern pupfish lived in a huge lake (Lake Manly) during the ice ages. When the lake dried up in the last 10,000 years, populations of pupfish were separated, and some still survive in isolated fragments of habitat. Some haven't made it, such as the Tecopa pupfish, a subspecies of *Cyprinodon nevadensis* known only from the Tecopa hot springs in the eastern Mojave, which was last seen in 1982. It is extinct.

for the Klamath and Modoc Tribes, these fish are a culturally important food source, and their life cycle influences ritual life. For years, the Klamath and Modoc Peoples have been advocating for the recovery of the suckers, seeking assistance in revitalizing these culturally significant fish, yet progress has yet to be achieved.

Turtles and Tortoises

Turtles and tortoises have a unique evolutionary history compared with other reptilian taxa and the amphibians. Like other reptiles, they are ectothermic, but they are distinguished from other reptiles by their shell. It is composed of bone with layers of keratin on top, and it provides them with protection from predators and serves as a bony framework for their body.

Desert tortoises are completely terrestrial, sea turtles live most of their lives in the ocean but must come to beaches to lay their eggs, and western pond turtles are primarily aquatic but go on occasional walkabouts to meet mates and then lay their eggs in burrows on land. While amphibians undergo a metamorphosis from aquatic larvae to terrestrial adults, tortoises and turtles hatch from eggs as miniature copies of the adult form. Tortoises and turtles lay soft-shelled eggs like most snakes and lizards (though a few snakes and lizards give birth to live young).

Desert Tortoise

The desert tortoise (*Gopherus agassizii*), now federally listed as threatened, holds a special place in the hearts of desert wildlife enthusiasts. These ecosystem architects shape their desert surroundings in ways crucial for the survival of numerous species. Utilizing their powerful legs, they excavate burrows that provide refuge from extreme temperatures, help occupants conserve water, and safeguard their eggs. Remarkably, a single tortoise can create as many as 12 burrows within its territory. These burrows serve as essential shelters for a diverse array of animals, including desert woodrats (*Neotoma lepida*) and burrowing owls (*Athene cunicularia*).

Power companies now use the Mojave Desert as a location for solar arrays, and vast areas have been fenced in and covered with solar panels. While alternative energy is needed, this type of development denudes the landscape of native vegetation, blocks wildlife movement, and has some impacts on all wildlife habitat on the site. Planning and coordination between government agencies, interest groups, and conservation alliances has produced the Desert Renewable Energy Conservation Plan. It designates specific areas for energy development but also protects critical habitat, prevents fragmentation, maintains wildlife corridors, and includes monitoring and adaptive management strategies for ongoing protection of tortoise populations.

There are only a few kinds of native turtles and tortoises of California: five sea turtle species, the western pond turtle, and the desert tortoise (*Gopherus agassizii*). Pond turtles are common and easily observed as they bask on logs or rocks near fresh water, from the Santa Ana River of the LA basin to the Klamath River of far Northern California. You have to be far luckier to encounter a desert tortoise, as they are shy and secretive. The best way to see sea turtles is to visit a nesting beach, none of which remain in California, though many are well known along both coasts of Central America. You can also view green sea turtles in the mouth of the San Gabriel River in Long Beach, where a power plant keeps the water warm all year long and the turtles like to relax in the warm water.

Amphibians

Amphi means "both," and *-bio* (*-bian*) means "life," so *amphibian* means "both lives"; amphibians are animals that live their lives both on land and in water. Many amphibians spend much of their lives "on land" (that is, in nonaquatic, terrestrial habitats) but "return" to water to breed. Because amphibians complete important life cycle events both on land and in the water and have highly permeable skin, they often are among the first vertebrates affected by environmental changes and pollution.

The Pacific chorus frog (*Pseudacris regilla*) is a great example. These frogs spend much of their time hunting or estivating far from water, but when it is time to breed, the males and females meet in ponds, lakes, and streams where they court, mate, and deposit eggs to develop in the aquatic environment. Chorus frogs would be unable to breed without puddles, ponds, and streams. When the eggs hatch, the resulting tadpoles require fresh water in which to swim, scrape algae from plant stems for sustenance, and hide from predators. Thus, important portions of amphibian life cycles may take place in both aquatic and terrestrial habitats.

Not all amphibians are so closely tied to ponds, lakes, and streams for reproduction. Many woodland salamanders have cut the tie to breeding in aquatic habitats and lay egg masses in moist cavities in logs or under rocks. This adaptation probably developed as a result of another widespread amphibian adaptation: small home ranges. Many small, slow-moving salamanders spend their entire lives within an area of just a few square yards. The small, black slender salamander (*Batrachoseps*) you may find under a log or garbage can on a wet winter morning may never move more than 15 feet from that spot during its entire life! Many patches of potential habitat are far from water, so some amphibians evolved ways to cut their ties to water in order to exploit those patches.

One of the important ecological characteristics of amphibians is soft, moist, water-permeable skin. Because amphibians lose water through their skin at about the same rate that water evaporates from a glass, they are in extreme danger of drying to death. Amphibians lose water through their skin far faster than other terrestrial vertebrates like snakes, lizards, birds, and mammals with relatively water retentive-skin. For this reason, amphibians have a suite of behavioral and life history traits that protect them from drying up. Often, they are active only during the wet season, and they sit out the dry season in a protected, cool, moist microsite, like 2 feet underground in a gopher burrow. Typically, they are active aboveground only at night, or during a rainstorm. One of the more fantastic adaptations along this line is "drinking by the seat of your pants." Western toads (*Anaxyrus boreas*), which live in relatively hot, dry habitats and are active at night during the dry season, have the

California Red-legged Frog

The red-legged frog is one of California's largest frogs, and one of the most terrestrial. Although red-legged frogs breed in ponds, bogs, lakes, swamps, and slow-moving streams, they spend much of their time wandering through forests. And yet, males can call to females at breeding time from 2 to 3 feet underwater! Red-legged frogs are slow growing, taking 3 to 4 years to mature. Once abundant throughout California, they now exist in a patchwork of isolated populations. They can be found in the foothills of the Sierra Nevada, in the Coast Ranges, and in pockets in Riverside and San Diego Counties. Named for their distinctive crimson underbelly and hind legs, these beautiful native frogs have suffered from habitat loss due to agricultural and urban development coupled with predation by nonnative species, especially bullfrogs. These impacts and resulting declines in the size of their populations and overall distribution warranted their listing as a threatened species in 1996. Local efforts at bullfrog control and red-legged frog recovery have been successful. For example, beginning in 2005, the National Park Service initiated bullfrog removal efforts in Yosemite Valley. Over the subsequent years, concerted efforts led to the successful eradication of the bullfrog on a landscape scale. This allowed for the introduction and restoration of California red-legged frogs, demonstrating the potential for effective invasive species management and amphibian recovery.

ability to expose a patch of thin skin between their hind legs (the "seat patch" or "pelvic patch") to bits of moisture, like dew on grass at dawn or damp soil. They absorb moisture through this "seat patch" in place of directly drinking.

Amphibian defensive chemistry is highly developed. Western toads and northwestern salamanders (*Ambystoma gracile*) have large, swollen areas on their skin (parotid glands) that hold defensive chemicals that can be excreted over their bodies as a milky ooze when threatened. These defensive chemicals may kill small predators (snakes or shrews) or sicken larger ones (raccoons or skunks). An interesting human application of amphibian defensive chemistry is the use of parotid gland secretions from the Sonoran desert toad (*Bufo alvarius*), which contain hallucinogenic tryptamines. People have been known to smoke toad as a way of altering their perception of reality! Perhaps the salamander most familiar to many Californians is the newt (*Taricha*), whose skin contains one of the most potent neurotoxins known, tetrodotoxin. The toxin is actually synthesized by a bacterium living on the skin of the salamander, not by the salamander itself, but nevertheless the animal is rendered deadly poisonous. At least one animal, the western terrestrial garter snake (*Thamnophis elegans*), has evolutionarily "learned" to detoxify the tetrodotoxin and preys on newts throughout much of the newts' range.

Dinosaurs

Many naturalists develop an early fascination with natural history sparked by their childhood curiosity about dinosaurs. Dinosaur bones are very rarely found in California because California was covered by ocean when dinosaurs roamed the Earth (in the Triassic and Jurassic periods of the Mesozoic era). Dinosaurs raise a topic every naturalist must come to terms with: extinction. While it is true that around 99.9 percent of all species that have ever existed on Earth have gone extinct at some point in time, it's also important to know that extinction rates today, accelerated by human-caused impacts, are up to 100 times higher than was typical before modern humans evolved. No longer can we assume that species will persist for a million years—the average species' time span during the nearly 4-billion-year history of life. Dinosaurs roamed the Earth, and in some ways dominated the terrestrial sphere, for about 80 million years. Then, about 66 mya they all suddenly died out—very suddenly—when a meteor 6 miles in diameter hit the Earth in the shallow Cretaceous seas off the coast of Mexico near the Yucatán Peninsula. This impact set off a chain of events that wiped out about 75 percent of the species that existed on Earth at the time, including most of the dinosaurs. This extinction event is perhaps the best known of the many major extinction events that have interrupted the course of life on this planet, but certainly not the most dramatic. That title probably belongs to the Permian-Triassic extinction event of 252 mya when about 95 percent of all species on Earth were wiped out in a tremendous spasm of global warming.

    Extinct means those species are gone for good—it's like saying good-bye forever, marking the end of their evolutionary journey. *Extirpation*, on the other hand, means a species no longer persists in a certain area or part of its original distribution. All of the grizzly bears in California were killed, but the species is not extinct. Grizzlies have been extirpated from California but still live in Montana, Idaho, Canada, and Alaska. Since the arrival of people in the New World, the list of extinctions is long. Many are prominent and well known. The mammoth, the short-faced bear, and the giant ground sloth are some of California's charismatic megafaunal extinctions. Many more are unremembered. When did you last see a fossil or poster of the Law's diving goose, whose most recent remains from Ventura County dated to 770 to 400 BCE? Ever seen one of your friends sport a hat with the giant island deer mouse on it?

The list of vertebrate animals that we know have gone extinct in California in the past 200 years includes the Tecopa pupfish subspecies (1982), Santa Barbara song sparrow subspecies (~1965), San Clemente wren subspecies (1968), thicktail chub (1957), Clear Lake splittail (1970), sooty crayfish (nineteenth century), and Pasadena freshwater shrimp (1933). Many insects, spiders, and mites have also certainly gone extinct without documentation. With around 300 species listed under the California Endangered Species Act, we are facing an extinction debt: species are temporarily holding on after habitat modification and loss, only to eventually go extinct if no action is taken.

Birds

While amphibians have moist, porous skin, reptiles have scaly skin, and mammals have hair growing from their skin, birds instead have feathers growing from their skin. These huge generalities that many schoolchildren know are actually good markers for some of the evolutionary and ecological distinctions of the major groups of vertebrates. Birds are the descendants of dinosaurs. They are characterized by the possession of feathers, an evolutionary novelty no other group of organisms has invented. Birds use feathers to stay warm or dry (insulation), to advertise to potential mates, to blend into their environment (camouflage), and perhaps most distinctly, to fly. Flight is so central to birds' fundamental nature that many other physiological structures can be traced to the importance of flight. For example, birds have more neck vertebrae than mammals, which allows them to scratch or preen their bodies with their bill. Birds also have a lightweight bill with a bony core and a cover of keratin but no teeth, to reduce their overall weight. Laying external eggs rather than giving birth to live young allows them to quickly get rid of weight. What do all these adaptations tell us about what is important for most birds?

Flight explains much about bird behavior. Birds fly to escape terrestrial predators like coyotes and snakes. Birds fly to avoid the unfavorable season in temperate latitudes (we call this migration). Birds are able to nest high in trees, on vertical cliff walls, on nearshore rocks, and inside hollow dead snags. And yet turkey vultures choose to nest on the ground, often at the base of a hollow tree trunk. Kingfishers and bank swallows dig tunnels in vertical dirt walls, and many birds, for instance bushtits, weave bags or "socks" in which to nest. Some birds, such as penguins, ostriches, and kiwis, have evolved other ways to get around. The variety of strategies that the 10,000 or so species of birds found worldwide have evolved to cope with the opportunities and challenges offered by the environment are nearly endless.

A second novel adaptation of birds is the use of song to communicate. The syrinx in a bird's throat is a unique, derived characteristic, not shared with other groups of vertebrates. It both helps to define birds and gives them the ability to sing. Birds sing or call to attract mates, to define and defend territory, to warn of the presence of predators, and to promote cohesion in social groups, like foraging flocks and family groups.

Most songbirds learn their songs, but for some birds, such as the tyrant flycatchers in North America, their songs are innate. For birds that learn their songs, the kind of "teacher" they have is terrifically important. If a band-tailed pigeon is raised in captivity by mourning dove parents, it will sing the song of the mourning dove. Separate populations of birds of the same species will sing different dialects of the same song. The white-crowned sparrows of the San Francisco Bay Area are well studied in this regard, with the dialect of birds born in San Francisco differing from the dialects of those born in Berkeley or Marin County. Another nuance in birdsong is that the same bird will sing slightly differently at different times of year, and at different times of day. When the red-breasted nuthatches begin to sing in the spring, they often sound like they don't know what a red-breasted nuthatch is supposed to sound like. It takes them a week or two of trying to get it right. And no two robins' songs are exactly alike. Every robin song is new and live, and it may be different from any other. Many brilliant people have spent the majority of their lives learning about birdsong and not gotten to the end of it.

Lizards and Snakes

It is easy to take living on land pretty much for granted, but life on land is the great evolutionary feat that led to much of the diversity of life on Earth today. The great innovations that made life on land possible are related to surviving hot, dry conditions. The first tetrapods on land were the amphibians, which are constrained by the need of moisture to reproduce. The

Acorn Woodpeckers

Acorn woodpeckers (*Melanerpes formicivorus*) at the nest. Photo by Bruce Lyon.

With its vibrant plumage and distinctive "laughing" call echoing through the forest, the acorn woodpecker captivates all who encounter it. But beyond its striking appearance and vocal prowess lies a fascinating cultural connection. Indigenous basket weavers skillfully incorporate feathers from these woodpeckers and other California birds into their baskets. The baskets contain meaningful woven patterns, are decorative, and hold ceremonial significance. They symbolize reverence for nature's bounty and the interconnectedness of all living beings.

The acorn woodpecker's survival is intimately tied to mixed oak woodland habitats. Its diet is based on acorns. Interestingly, the acorn woodpecker doesn't immediately consume the acorns it collects, but works cooperatively with its family group to store acorns in granaries. The birds spend days drilling holes in snags, dead limbs, utility poles and buildings, then poking acorns into the holes for later consumption, thus ensuring a vital food source for later in the winter. Acorn woodpeckers are cooperative breeders, living in large family groups in which several birds, almost always related, may become parents and defend territories containing their precious acorn granaries. It is great fun to watch these family groups interact. An acorn woodpecker family may have as many as 7 breeding males, 3 breeding females, and 10 helper birds in a single group. Conserving oaks and acorn woodpeckers is an ecological imperative and a cultural responsibility.

great innovation of further vertebrates was the amniotic egg, which is porous to gases but retains water, thus allowing birds, crocodiles, turtles, lizards, snakes, and mammals to invade increasingly dry habitats. Another pivotal adaptation lizards and snakes have developed is dry, scale-covered skin that retards moisture loss. Lizards and snakes are unlike birds, crocodiles, and turtles in that lizards and snakes separate the water from their waste products to avoid losing it in disposing of the waste.

Another huge difference between lizards and snakes versus mammals and birds is their metabolic rates. In order for birds and mammals to be fast and active, they must have high metabolic rates. For birds to fly, they must beat their wings rapidly and continuously, thus producing an extreme oxygen demand. Compare this with a rattlesnake that may lie without moving for days, waiting for an unsuspecting rabbit to pass within striking distance. The energy demands are very different. Don't be tempted to think of lizards and snakes as inferior or primitive because they are ectothermic or poikilothermic. Ectotherms are actually, in some ways, at an advantage with their "undemanding" metabolism. Birds and mammals use about 30 to 80 percent (in colder climates) of the energy they consume to maintain their body temperature. So lizards and snakes can survive on about one-tenth the food birds and mammals need. When food is scarce or unavailable, lizards and snakes can wait for abundance to return, whereas birds have to fly to Central America. In many habitats the biomass of lizards and snakes combined is larger than that of birds and mammals combined, in part because their smaller metabolic demands require less energy from the ecosystem. Snakes exhibit a suite of traits that distinguish them from other vertebrates and characterize them ecologically: they are extraordinarily mysterious, eat infrequent but large meals, and depend on smell rather than sight or sound for most of their information about the world. Snakes are typically limbless. Snakes and lizards also have paired penises (hemipenes), a derived adaptation shared with no other group of vertebrates. Both lizards and snakes shed their skin to renew it, in marked contrast to birds, which regrow individual feathers, and mammals, which regrow individual hairs. Most lizards shed their skin irregularly, dropping small, ragged patches over the course of a week or two. Snakes (and alligator lizards) loosen their skin as a whole package, break it loose at the mouth, and then crawl out in as little as 5 minutes, leaving the inside-out skin with the tail pointing in the direction the shedder departed.

A few snakes (boas, pythons, pit vipers) use heat (infrared radiation) to gain information about the world. A pit viper (such as a rattlesnake) may actually form an image in its brain based on infrared radiation.

Pacific gopher snake (*Pituophis catenifer*). Photo courtesy of Greg Damron.

Perhaps the most important attribute of snakes is their devoted use of chemical cues to navigate and negotiate their worlds. Although snakes have nostrils for breathing, they rely on their forked tongue for chemical cues from the environment. After flicking its tongue, a snake retracts it and transfers collected chemicals from the forked tips, called tines, into two sensitive pockets on the roof of its mouth. The right tine contacts the right pit, and the left tine the left pit. This allows the snake to perceive concentration differences from side to side, thereby allowing it to follow the stronger scent trail to a prey item or potential mate.

Lizards are different from snakes, but they both have eggs that are soft and leathery, unlike the hard, brittle eggs of birds. In California, lizards are desert specialists, with far more species and greater numbers of individuals in hot, dry habitats than in cool, moist forests. One of the unique attributes of most lizards is the ability to lose and regrow their tails. Lizards do this so commonly that they have special joints in the tail vertebrae to facilitate tail loss. Lizards have a special mechanism for stopping the flow of blood at the lost joint. The tail, meanwhile, may wriggle frantically when separated from the body, distracting the predator from the other potential food item nearby. Lizards regrow their tails over time. The tail loss tendency seems to be an adaptation to high rates of predation: better to lose a tail than die! Many lizards establish and defend territories, marking them with chemicals produced by scent glands located on the legs. The territory is commonly defended with ritualized threat displays that involve doing push-ups on prominent perches to show off brightly colored belly scales that are normally invisible when the lizard is not displaying.

Mammals

People have a love-hate relationship with animals. On the one hand, people love their pets, with about half of US households having a dog and nearly as many a cat, but most folks fear rats. Tourists pay good money for the chance to see megavertebrates such as whales or bears and value that experience highly, but they usually ignore smaller, more common mammals such as squirrels. And how many of us take the time to watch a spider weaving a web on a bush in the garden or are willing to share our living space with dangerous animals? The current political dialogues surrounding resurgent mountain lion populations and the return of wolves to California are examples of the difficulty people have allowing dangerous animals to coexist with us. The fear and the facts do not always match up: though numerous people consider wolves scary predators, the likelihood of wolves killing people is close to zero.

Mammals differ from other vertebrates by having skin with hair, by nourishing their young with milk produced in mammary glands, and by having diverse tooth types. Like feathers on birds, hairs are a mammalian invention. Hairs are used to keep warm and cool, to attract a mate, and for camouflage, much as are the feathers of birds. Even marine mammals (whales, dolphins, seals, otters) have at least some hair at some point in their life cycle. Many, perhaps most, mammals also use hairs for sensory purposes: they feel with them. The hairs about the face normally called whiskers on cats and dogs are actually long, stiff, sensory hairs. Whiskers can tell a bobcat that it is about to touch a tree or a spiny blackberry stem before the animal bumps into those things with more sensitive features like its nose or mouth.

Most people will be familiar with the idea of lactation: feeding offspring with a rich, sweet, liquid food for the first days or months of life. This is another typically mammalian trait. Even whales and dolphins, which nurse their young underwater, rely on this method.

Teeth in most nonmammalian vertebrates are relatively uniform. Although snake teeth and lizard teeth may look quite different from

Bats

A pallid bat (*Antrozous pallidus*) carrying a long-horned grasshopper. These bats catch prey almost exclusively from the ground or foliage, mainly feeding on large insects, scorpions, and centipedes, and they pollinate cactus and agave flowers. © MerlinTuttle.org.

Bats provide a great example of how society's perceptions of wildlife can change. In the 1950s and 1960s, bats were commonly viewed in America as pests or as dangerous carriers of rabies. This is factually inaccurate; bats almost never transmit rabies to people. Far from being pests, they are among the most helpful, efficient protectors of our health and agricultural production. A single bat may eat 1,200 insects in one hour! Without bats we would be overwhelmed by mosquitoes, which, unlike bats, are serious, dangerous transmitters of disease to humans. Bats also protect our agricultural crops by gobbling down the insects that would otherwise consume them. Since about 1975, there has been a huge reversal of the image of bats in the American public mind. People now fly across the country for the purpose of watching bats fly out of their daytime roosts at sunset, a knockout sight.

each other, all the teeth in an individual lizard's or snake's mouth are basically the same. Mammals have diversified teeth for different functions: the grinding teeth of a grazer like a deer look very different from the tearing teeth of a predator like a coyote. One mammal may even have two or three very different tooth styles: the look and function of the beaver's chisel-like front teeth are very different from those of its rear grinding teeth. The tooth specializations of mammals represent specific evolutionary adaptations, which are central to the habitats and niches mammals have been able to occupy.

There are three main evolutionary groups of mammals: monotremes, marsupials, and placentals. The monotremes stretch the idea of what a mammal is, since they give birth by laying eggs! There are two extant genera of monotremes, the spiny echidna of New Guinea and the duck-billed platypus of Australia. In monotremes, the eggs are placed in a marsupial-like pouch where they hatch and the infants are fed milk. Monotremes further distinguish themselves from other mammals by lack of teeth! Monotremes are a good illustration of the idea that for every generality about how the world works, nature has produced an exception.

Marsupials should be familiar to many through popular culture: just picture a mother kangaroo with a joey looking out of her pouch. Marsupials reached their greatest diversity in Australia, New Zealand, and New Guinea, where different lineages of marsupials occupied all of the ecological niches of mammalian browsers, grazers, predators, and burrowers. The one marsupial we commonly see in California is the Virginia opossum, whose ancestors came from South America during the Great American Interchange, which occurred after the formation of the Isthmus of Panama about 3 mya. This species is continuing to move north, in part because it is well adapted to urban environments and also as global temperatures rise. All marsupial young are born very early in development and migrate to a pouch, also called a marsupium, where they are fed milk and grow.

The third and final group of mammals is the placentals, by far the most diverse lineage of mammals and the dominant group of vertebrates on the planet today. In placental mammals, the young undergo considerable development within the body of the mother. They are nourished by a specialized organ, the placenta, which allows nutrients, oxygen, and waste products to pass between mother and offspring.

HUMAN ACTIVITY AND DOMESTIC AND INTRODUCED ANIMALS

We share the Earth with animals as pets, neighbors, and competitors. Human activity influences wildlife populations in two ways: through changes in land use and through the introduction of domesticated and

Grizzly Bears

The California grizzly bear is a distant memory featured on the state flag. Indigenous Peoples had long coexisted with over 10,000 of these bears, from the Sierra Nevada to the Coast Ranges, and viewed them as an essential part of the community of life. For many Tribal communities, grizzly bears are one of the original teachers and are sacred animals. There were traditional taboos on harvesting the bears or eating their meat in some places.

Colonial settlers hunted and trapped the grizzlies, resulting in their extirpation from California. The grizzly bear was a keystone species, with important roles such as regulating prey populations and dispersing seeds, and its absence has profoundly impacted California's ecosystems. With only about 2,000 grizzly bears in the lower 48 states, their future remains uncertain.

In 2016, nearly 100 years since the last wild grizzly sighting in California, a group of scientists, leaders from California Tribes, and advocates from conservation nonprofits formed the California Grizzly Alliance to bring the bears back.

other animals. Changes in land use alter habitat by converting areas to development, agricultural, and other uses or through habitat modification and fragmentation. All of these activities favor some wildlife species over others. Species that favor open spaces and edge habitat tend to thrive, while species that require lots of cover, cooler temperatures, or continuous habitat tend to suffer.

Some fauna were introduced intentionally as work animals or pets. Some were introduced accidentally as hitchhikers in ships or cargo. Many introduced animals live with relatively little impact on the environment, while others compete so strongly with native species that they eliminate them and can cause extirpation. One example is the bullfrog. Bullfrogs were brought to California as a food source but made their way into wildlands, where they quickly spread. Bullfrogs prey upon native species of insects, frogs, turtles, and fish. In locations throughout California and much of the western United States, bullfrogs have led to the decline or displacement of native amphibians. Invasive species do not always originate from distant places. For example, the Sacramento pikeminnow is native to the Sacramento area but has become invasive in the nearby Eel River, where it has impacted populations of salmon and steelhead.

Domesticated animals, when not managed properly, can also have a deleterious effect on native species and the environment. One fairly well-known example is the effect of cats on songbirds, with free-ranging domestic cats killing an estimated 1.3 to 4.0 billion birds and 6.3 to 22.3 billion mammals globally each year. Some domesticated animals have become well established or naturalized in the wild. An example is feral pigs, now common in California's woodlands, which are a concern for land managers because of the damage their rooting can do to sensitive or restored habitats and because of the tons of acorns they eat. Acorns are the prime food resource for a long list of California animals, and every pound of acorns pigs consume is a pound that gray squirrels, bears, deer, and acorn woodpeckers will not use to make babies. Another problem is that wild pigs are nonnative predators that can impact native prey species. In fact, 40 percent of the pigs studied in the Diablo Range had native vertebrate prey in their stomachs, demonstrating that the impact of pigs extends to small mammal and other vertebrate consumption.

Saving species can be challenging, with the threats domestic and invasive species present, plus the need to protect more habitat and prevent rapid climate change. And yet the rewards of helping animals survive are clear. Every animal has a story to tell and a lesson to teach. Give yourself time to observe the meticulous habits of bees or the patterns of a butterfly fluttering from one flower to another. By watching and learning more about animals, you're not just gaining knowledge; you're embarking on a journey of wonder and connection to the natural world. Whether you're in the desert or a dark forest, animals likely smell or see you, and in time you will have the thrill of observing some of them. Stay still and let the animal kingdom inspire and amaze you!

Explore!

VISIT A MIGRATORY PATHWAY STOPOVER

As an important stretch of the Pacific Flyway, California is one of the best places in the country to watch migrating birds. Great places include the Marin Headlands, the islands in Channel Islands National Park (take a boat out of Ventura), and Monterey Bay. Other ideas include an Audubon sanctuary, a wildlife refuge, an estuary, or even the local sewage treatment plant ponds.

Monarch butterflies migrate to California overwinter. Although monarch numbers are declining, you can still find winter aggregations at Natural Bridges State Beach in Santa Cruz, Monarch Grove Sanctuary in Pacific Grove, Pismo Beach Monarch Grove, the Ellwood Main Monarch Grove and the Coronado Butterfly Preserve in Goleta, and Camino Real Park in Ventura.

VISIT A WILDLIFE REFUGE, NATURE CENTER, OR NATURAL HISTORY MUSEUM

The best way to hone naturalist skills is to take a field guide, some binoculars, and a notebook and head out to a wildlife refuge. A list of refugees nationwide can be found on the US Fish and Wildlife, National Wildlife Refuge System website.

Also visit one of California's many nature centers or natural history museums, such as the California Academy of Sciences in San Francisco, the Pacific Grove Natural History Museum, the Santa Barbara Museum of Natural History, the Natural History Museum of Los Angeles County and the La Brea Tar Pits, and the San Diego Natural History Museum. California also has many outstanding local nature centers.

GO TO A NATIVE PLANT GARDEN AND FOLLOW THE POLLINATORS

First, try to identify as many plants as you can. Then find a native bee and see what plants it pollinates. The California Native Plant Society has a list of botanical gardens with extensive collections of native plants on their website.

BIRD-WATCHING

A great way to become familiar with birds, and nature observation in general, is to watch birds nesting near your home or drinking from a birdbath. Start by sitting in a green space for 5 minutes and focusing on the sounds around you. Then spend another 5 minutes without

(continued)

moving, still listening intently, and see if you can locate any of the birds you heard. If you repeat this exercise a couple of times each week or daily, within a year you will come to know the birds near you.

PARTICIPATE IN THE CHRISTMAS BIRD CENSUS

Every December, the National Audubon Society engages thousands of volunteers all over the country to conduct a bird census in their local area. To get involved, contact your local Audubon chapter.

TRACK AN ANIMAL AND IDENTIFY SCAT

Learn the tracks and scat of common mammals and birds in your area, and try to identify them as you hike. Keep a journal of what you find, and draw tracks you come across.

LEARN ABOUT ALTERNATIVES FOR PEST CONTROL

Most pest control techniques involve the use of poisons to kill the un-desired creatures. Those poisons have the unfortunate tendency to end up in the water system and often don't solve the problem. Natural pest control can be used to solve the problem at its root. (Is there a hole that needs to be plugged? Standing or leaking water?) And it utilizes less-toxic or nontoxic methods. You can learn more at the University of California's Integrated Pest Management website.

7

Energy and Environmental Issues

Most of us live in a world of convenience. Most of us don't have to contend with 110 degrees or 17 degrees. But by living and working indoors, we can become detached from the environments we live in. And what is the cost of all this comfort? Not the actual cost of the monthly HVAC bill, but the real cost that includes losing connections with the myriad of species living outdoors. To be a naturalist is to seek connections and ask questions, questions that may not have easy answers. Does around-the-clock electricity availability mean losing our ability to see the stars? Does our daily commute mean filling the atmosphere with carbon dioxide and causing sea level rise? Does seeking more solar power mean covering the desert in solar panels and possibly driving the magnificent desert tortoise to extinction? This chapter, like all of the others in this book, is about connections. But in this chapter, you will be faced with some of the most uncomfortable connections naturalists, and indeed everyone, must address. Take it slowly, don't let it overwhelm you. Think about the connections that give you strength, resilience, and joy.

SOURCES OF ENERGY ON EARTH

Energy is a fundamental unifying concept for physical and biological scientists because it drives all Earth systems. Energy is neither created nor destroyed; it simply changes form. Therefore, energy can never be "used up." When nuclear fusion in the dense gas cloud of the sun releases the energy of the atom, it transforms that energy to various forms of electromagnetic energy. The exact amount of energy that was held in the nuclear bonds of the hydrogen or helium atoms reacting in the sun is the amount released as light, heat, and other forms of energy.

Energy on Earth is derived from two sources: geothermal heat generated within the core of the planet, and electromagnetic energy coming from the sun, most notably heat and light.

Energy within the Earth

Earth's interior is not a simple homogeneous mass of solid rock. There is quite a lot of variation among layers, in terms of composition, density, and phase (liquid vs solid). The Earth's core is a tremendous source of heat—some theorize that the Earth's core is hotter than the surface of the sun, around 9,932 degrees! This intense heat is derived primarily from two sources, primordial heat and radioactive decay.

Primordial Heat

The Earth was first formed 4.5 billion years ago by bits of rock and dust coalescing as they orbited the sun, and these bits of rock contained a certain amount of energy. As the rocks piled up deeper and deeper, some of that energy was trapped. Some of the kinetic energy of the rocks' motion was also converted to thermal energy (heat), and thus huge amounts of energy were trapped deep in the core of the developing Earth as heat. This heat, coupled with heat generated by radioactive decay in the core, has been enough to keep the outer core in a liquid state, despite the great pressure the material is under. Heat conduction from the outer core drives the convective flow of the solid but ductile mantle, and ultimately plate tectonics, on the surface of the Earth.

Radioactive Decay

The heat produced by radioactive decay of, for example, potassium-40, uranium-238, uranium-235, and thorium-232 combined with the original heat from the Earth's formation drives the global geological cycle, including the motions of tectonic plates and the continents that ride as components of those plates, the eruption of volcanoes, the thundering

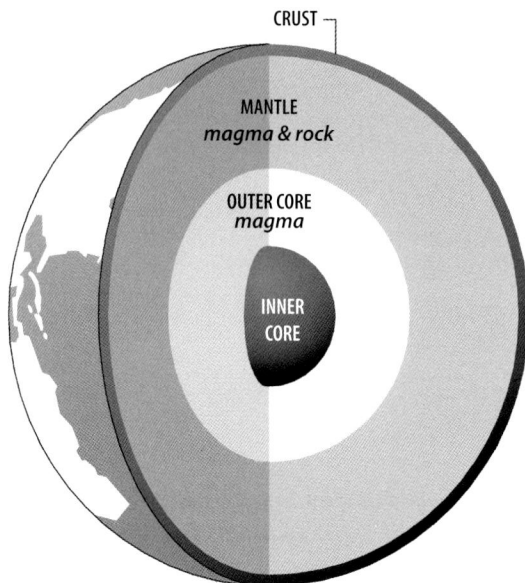

Internal structure of the Earth. Not to scale; the crust should be far thinner but is drawn so that it will show. Used with permission from the National Energy Education Development Project, www.need.org.

of earthquakes, and the uplifting of mountains. The Earth is constantly releasing the heat stored and produced deep within its core. The moon and Mars are geologically dead because they have lost all of their internal heat. Thanks to the insulating properties of the crust, the Earth will probably retain its core heat until long after the sun swells to swallow it. If the Earth's internal heat were all dissipated, like the moon's has been, the motions of the continents would end. Volcanoes would no longer erupt and add to the land surface area, and the continents would eventually be washed away by erosion into the oceans and not be rebuilt.

Energy from the Sun

In order to carry out any activity—waking up, walking, breathing, eating, flying, hunting, mating—energy is required. Virtually all of the energy driving the common biological world is derived from nuclear reactions taking place within the sun. The sun is composed of hydrogen and helium, which are involved in continuous nuclear reactions. Nuclear fusion means the joining of two atoms, while nuclear fission is the splitting of the atomic nucleus. Both fusion and fission liberate huge amounts of energy. The sun is a giant ball of gases undergoing continuous fusion reactions. The nuclear reactions of the sun send vast amounts of energy out into space as electromagnetic waves, which are captured by photosynthetic organisms and drive all life on Earth.

THE ENERGETIC BASIS OF LIFE

When the sun's energy arrives on the Earth, it is transformed in two key ways. Some of the solar energy is used to change water from solid to liquid or from liquid to gas, a transformation that requires large amounts of energy. In fact, water is one of the primary acceptors of energy from the sun. Without the sun's continuous energy input, all the water on Earth would gradually lose its energy to the dark void of space and would freeze. The various transformations of water from one phase to another drive the planetary climate system. The energy transformations of water from ice (solid) to river (liquid) to water vapor (a gas that condenses into clouds) to rain (liquid) or snow (solid) are examples of some of the primary determinants of the physical conditions we take for granted, and upon which life depends.

The second form of energy transformation on Earth is the change of the sun's electromagnetic radiation into the energy of chemical bonds in living systems, through photosynthesis. This is a fact that is easy to overlook, but without plants' ability to convert the sun's energy, there would be virtually no life on Earth. Green plants and algae use the sun's energy to build chemical compounds capable of storing that energy. The primary chemical compound that stores the sun's energy and powers life is glucose: $C_6H_{12}O_6$. Glucose is the transportable form of energy in biological systems. It is used to drive the process of aerobic respiration, where glucose reacts with oxygen to release energy, carbon dioxide, and water. Respiration provides energy for all the cell's activities. Glucose supplies the energy for growth, reproduction, and all other aspects of biological activity. In sum, the sun's energy is captured in the cells of green plants, and those plants then provide the chemical energy to drive almost all other forms of life on Earth. There are interesting exceptions to life relying on photosynthesis, such as complex communities found in

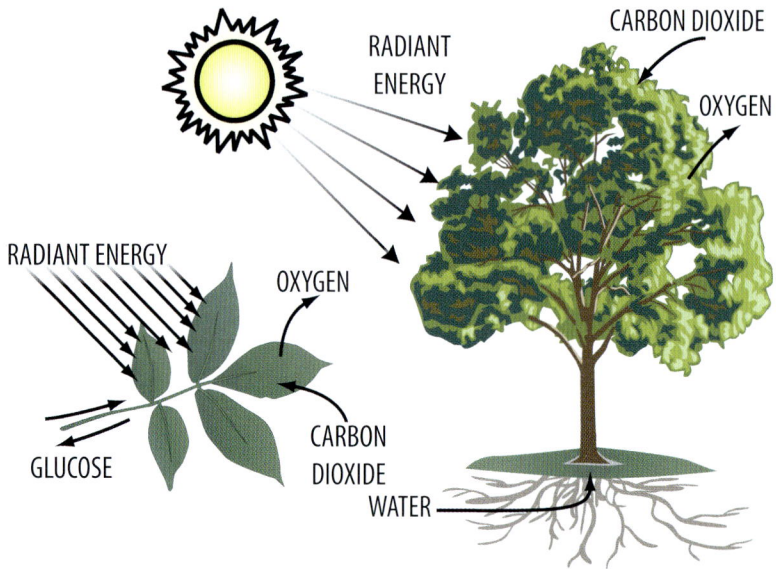

$$\text{water} + \text{carbon dioxide} + \text{sunlight} \longrightarrow \text{glucose} + \text{oxygen}$$

$$6H_2O + 6CO_2 + \text{radiant energy} \longrightarrow C_6H_{12}O_6 + 6O_2$$

Photosynthesis is the energetic basis of all life on Earth. Without green plants there would be no life as we know it. Used with permission from the National Energy Education Development Project, www.need.org.

deep-sea hydrothermal vents where no sunlight reaches. Life at these vents begins with the thermal energy being released from deep within the planet. Organisms there convert energy into food in a process known as chemosynthesis.

Energy transformations in biological systems are inefficient. Each time a rabbit eats some grass, only about 10 percent of the energy stored in the grass is converted to usable energy in the rabbit. When a rattle-snake eats the rabbit, about 10 percent of the energy is successfully captured by the snake. When a coyote or a red-tailed hawk then eats the snake, the energy "loss" is similarly about 90 percent. Each time energy is exchanged in a food web, the loss in conversion is about 90 percent. This is a limiting factor in the number of trophic levels a food web can support. It is the reason food chains almost never go beyond four or five transformations. It is more energetically efficient when people eat producers like grains, vegetables, seaweed, and fruits.

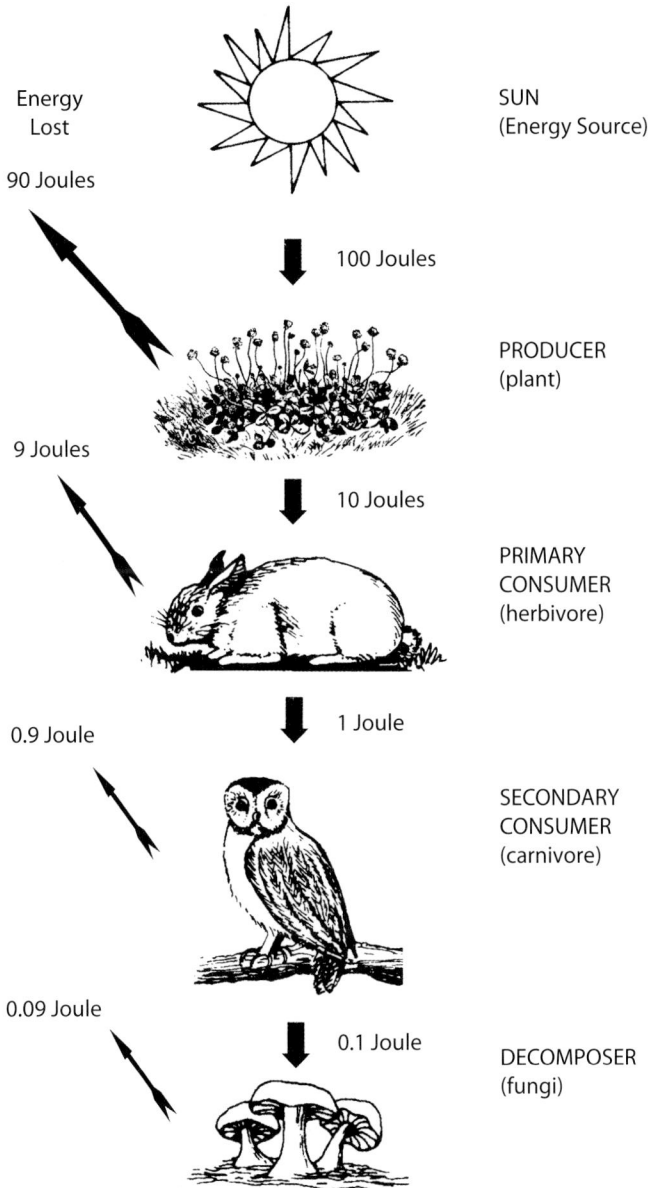

Energy Lost

90 Joules

100 Joules

9 Joules

10 Joules

0.9 Joule

1 Joule

0.09 Joule

0.1 Joule

SUN
(Energy Source)

PRODUCER
(plant)

PRIMARY
CONSUMER
(herbivore)

SECONDARY
CONSUMER
(carnivore)

DECOMPOSER
(fungi)

Energy loss in a food chain: 90 percent of the usable energy in an organism is lost at each step. The 90 percent is converted to heat and other unrecoverable forms of energy. Adapted from the University of Maine Cooperative Extension's *Connections to Our Earth: Leader's Guide*, Orono, Maine, 1995.

ENERGY USE BY PEOPLE

People are masters of turning the chemical bonds of matter into energy. Energy warms our homes. It powers cars, the trucks that bring food, and the tractors used to grow food. Electrical energy powers computers, radios, washing machines, and water heaters. Virtually everything about human life is intimately bound to energy use, though the dominant forms of energy have changed over the millennia.

The two most common forms of energy Americans are familiar with are gasoline and electricity. Most large-scale electricity generation is done using a turbine powered by wind, water, coal, or natural gas. The power sources for turning the turbine differ, but once a turbine is set in motion, the mechanical energy of the rotating shaft can be used to generate electricity. Gasoline isn't technically a form of energy, but a potential source of energy. Only when the gas is burned and the controlled explosion is converted to mechanical energy (the movement of the piston) can the original source (the gas) be used.

Fossil fuels are carbon-rich compounds formed from the fossilized bodies of ancient plants and animals. Ancient plants captured the sun's energy, some of this was transferred to ancient animals, and then both were buried for huge spans of time. Fossil fuels are the most concentrated and therefore the most convenient form of energy on Earth. Unfortunately, fossil fuels are not renewable on a timescale of interest to humans. There is only so much oil in the ground. For this reason, fossil fuels are termed nonrenewable resources. Unlike energy from the sun, which arrives on Earth at a constant rate essentially in perpetuity, fossil fuels are a one-time bonanza. Fossil fuels dominate world energy use. However, over the past decade, there has been a gradual decrease globally in the percentage of energy sourced from fossil fuels, down to about 80 percent as of 2022, even as total energy consumption has grown by about 2 percent per year.

Fossil fuels come in three general forms: coal, oil, and natural gas. They have, for about a century, been the main energy sources powering society. Fossil fuels are inexpensive to obtain and convenient to transport and store. The technology to transform fossil fuels to electricity is simple and well developed.

Coal is a solid and is extracted from the earth using heavy machinery. Coal is used for firing power plants to produce electricity, and for steel production. Oil is a liquid, procured by drilling and pumping, and is used to produce gasoline, diesel fuel, and plastic products. Natural gas (methane) is the gaseous form of high-energy fossil carbon and will be familiar to many as the cleanest-burning form of home heating energy.

California has no coal-fired power plants, and as of 2022, only a very small portion of the energy used in California comes from coal-generated electricity imported from adjoining states. This is commendable in terms of maintaining air quality. Coal is a dirty source of energy, and coal-fired power plants are notorious contributors to both global carbon issues and regional air quality issues. The disadvantage of using natural gas for producing electricity is that it is an inefficient use of fuel, compared with using it directly to heat homes. Only about half of the energy stored in a given volume of natural gas is captured as electricity in a power plant, whereas burning it to heat a home can be 96 percent efficient. In 2022, natural gas was the single largest source of California's electric power mix, 36 percent.

The extraction and use of coal, oil, and natural gas create undesirable by-products. Coal extraction (mining) is highly dangerous work and employs methods, such as mountaintop removal, with extreme environmental consequences. Oil extraction and transportation can produce accidents that seriously contaminate sensitive ecosystems. Most notoriously, extraction and combustion of the carbon-rich molecules that store energy in fossil fuels produce CO_2 (carbon dioxide), a greenhouse gas. Natural gas is methane (CH_4), itself a potent greenhouse gas. Both CO_2 and methane are primary contributors to climate change. Ethanol, used as an additive or an alternative to gasoline, burns more cleanly than gasoline in terms of particulates but similarly produces CO_2 emissions, which are undesirable in a warming world.

Another form of nonrenewable energy is nuclear power. The best estimates are that the current relatively modest rate of consumption of uranium-235 (U-235, the "fuel" used in nuclear reactors) could continue for between 100 and 200 years. If the rate of consumption of U-235 were dramatically increased by the building and operation of many more nuclear power plants, the global supply of U-235 would be depleted more quickly. Nuclear power is clean in the sense that operating the power plants does not emit carbon dioxide and thus does not contribute to global warming. However, like fossil fuels, nuclear power is a nonrenewable form of energy, and supplies of U-235 will eventually be exhausted.

More-immediate drawbacks to nuclear power are the cost; the highly dangerous, long-lived radioactive waste produced in the process; and the possibility of a nuclear meltdown. Nuclear waste may remain highly toxic for 600 years or more. Many engineers argue that the technical problems of storing dangerous nuclear waste are surmountable, but locating a storage area for the waste is usually blocked by political and social problems. (Who wants a nuclear waste site nearby?)

Another problem associated with nuclear power plants is the highly technical process of running them safely. When the process is not carefully monitored and controlled, the outcome can lead to events such as the meltdown at Chernobyl in 1986. In addition, events such as the combination of a giant earthquake and a tsunami can overwhelm even well-run plants, as was the case in Japan in 2011. Nonetheless, as a low-carbon, clean power source, nuclear energy is once again gaining traction because of its potential to help reach a net-zero carbon future.

Geothermal energy is an important part of the state's carbon-free energy grid. California is home to the world's largest geothermal field, The Geysers, located about an hour north of San Francisco. In 2022, California received 5 percent of its electricity from geothermal sources.

Geothermal energy is a clean, non-carbon-emitting power source that in some ways straddles the border between a renewable and a non-renewable energy source. The mass of heat trapped in the interior of the Earth would essentially be inexhaustible if it were all available to be tapped. However, the portion accessible from the surface is small and often distant from major metropolitan areas where the power is needed.

Geothermal has problems with disposal of contaminated water. Another challenge for geothermal energy is that the potential sources are sometimes located in some of the most pristine park lands (Yellowstone is an example), where people will object to industrial development. Geothermal energy can also become exhausted. That is, heat can be extracted faster than it is generated in the interior of the Earth, making geothermal a nonrenewable source, as has happened in New Zealand.

Wood for fuel also straddles the border between renewable and non-renewable. Wood is the primary cooking and heating fuel for much of the world. Wood can be readily depleted in some areas and overabundant in other areas. Open pile burning and older woodstoves can be polluting. However, new woodstoves designed to prevent greenhouse gas emissions, along with small power plants that utilize wood chips, offer innovative clean heating solutions in areas with an excess of wood. At the same time, we need to be cautious moving firewood to new locations, as it can unintentionally spread tree diseases like sudden oak death, as well as invasive insects like the goldspotted oak borer, which can have devastating impacts on ecosystems and forest health.

The renewable forms of energy available to humans include solar, wind, hydroelectric, wave, and tidal. While fossil fuels are concentrated forms of energy, renewables are dilute sources of energy.

Solar power has huge potential, for both small-scale and industrial applications. Solar energy can be utilized in many forms, from small-scale home water heaters, to industrial-strength solar arrays in the desert that use curved mirrors to heat fluids, which turn turbines to generate electricity. Houses and office buildings can be heated passively by orienting windows to capture solar energy. Another method of capturing the sun's energy directly is with solar cells, which capture the sun's electromagnetic energy and convert it to electricity.

Solar is a clean source of power that does not emit carbon during the operation phase. All sources of power emit carbon during the manufacturing phase (for example, in making solar panels and manufacturing turbines), thus the common term *carbon neutral* can be misleading. Nonetheless, solar is a renewable source of power, and the prognosis for the sun to rise tomorrow is good.

According to the International Energy Agency, demand for fossil fuels globally will likely peak in 2030 and be replaced by demand for renewable energy. Costs for producing renewable forms of energy have come down dramatically in recent years, and new electricity generation projects globally have been dominated by renewables.

California has set ambitious clean energy goals. In 2018, California established a landmark policy requiring 100 percent of the state's electricity to come from renewable energy and zero-carbon resources by 2045, with interim targets of 90 percent clean electricity by 2035, and 95 percent by 2040, and the state is making substantial progress toward that goal. According to data from the California Energy Commission, in 2023 approximately 39 percent of the state's electricity came from renewable sources such as solar, wind, small hydro, biomass, and geothermal, and that figure rises to nearly 58 percent if other non-fossil-fuel sources such as large hydro and nuclear are included. At the same time, energy use per capita has consistently fallen in the last few

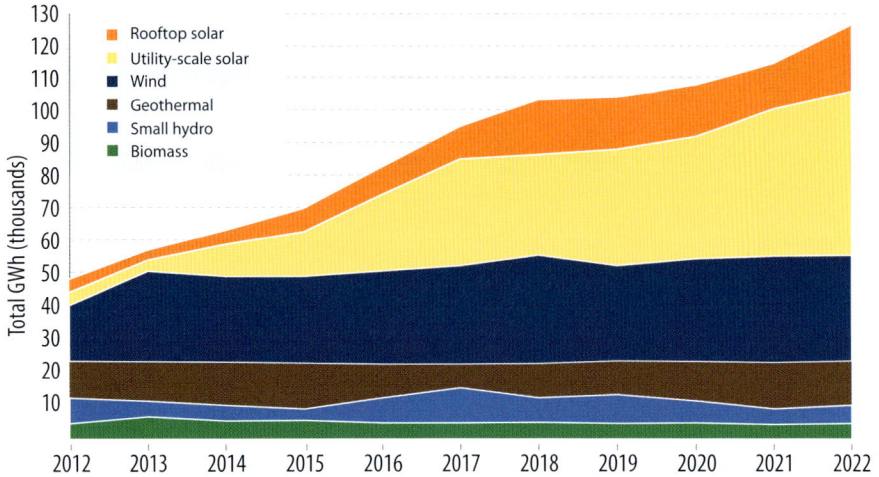

Renewable energy generation in California, in thousand gigawatt-hours, 2012–2022. California lawmakers have set a goal to reach 100 percent of the state's electricity generation from carbon-free sources by 2045. In 2022, about 36 percent of the power grid mix was from renewable sources. Courtesy of the California Energy Commission, 2023.

decades, and California uses less energy per capita than any other state. California is progressing toward a carbon-neutral future, having already achieved 100 days of energy supplied solely by renewable sources.

Solar is a leading energy source in new renewable energy projects, both in California and globally. However, utility-scale solar farms are land consumptive, and when that means covering extensive amounts of fragile and diverse desert ecosystems, this presents serious concerns for biodiversity conservation. An alternative approach for avoiding these impacts is to locate renewable energy projects on degraded lands such as former mine sites or above existing parking lots and on rooftops. When solar panels cover parking lots, they have the added benefit of shading blacktop and reducing the urban heat island effect. Solar panels are also now being developed to cover some of California's aqueducts, with the multibenefit goals of reducing evaporation and improving water quality while generating renewable energy.

Wind power captures the kinetic energy of wind via large arrays of wind turbines. Wind turns the turbines, and they rotate generators, which produce electricity. Wind is a clean, non-carbon-emitting source of power. The main limitation of wind power is the lack of places where the wind is strong and reliable enough to make operation economical. Wind farm proposals can run into local opposition based on aesthetics ("not in my backyard") and concerns about impacts on bats and migratory birds, while offshore projects can raise concerns about marine mammals. Significant efforts are being made to reduce or prevent impacts of wind energy generation on animals. For example, recent advances in wind farms have substantially reduced bird and bat fatalities. These advances include avian-friendly turbine designs, location selection to avoid bird migration routes and bat nesting areas, and equipment or procedures to stop blade movement during bird and bat activity. Avian radar systems can shut down turbines if birds are detected, and turbines can be turned off at dawn and dusk during known bat activity periods. Wind power generation is an important part of the clean energy portfolio and has increased dramatically in recent years, particularly in the Midwestern states. In California, wind energy represented about 11 percent of electricity generation in 2022.

Hydroelectric energy captures the mechanical energy of falling water via dams and turbines to generate electricity. The percentage of electricity from hydropower (small and large) in California, 10 percent, is higher than the national average. Hydropower is clean in the sense that it emits no carbon and does not contribute to global warming. The unfortunate side of hydropower is that damming rivers often destroys them as rivers. The Pacific salmon crisis, with 11 species of salmonids on the federal endangered species list and others sure to join it, is largely a result of hydroelectric dams. Damming of the Colorado River has caused similar problems. Before the dams were built, the Colorado River system was a seasonal, muddy, warm-water system. The dams have transformed the river into a cold, clear, low-nutrient system. Before the dams, the Colorado River supported 8 native species of fish and virtually no nonnative fish species. Currently, the Colorado supports about 20 species of nonnative fish, 3 of the original 8 native species of fish are extinct, and 2 are endangered as a result of the ecological changes caused by hydroelectric dams.

Ocean wave power and tidal power are largely ideas in the development stage. Both are essentially inexhaustible supplies of energy, if the technical challenges to tapping them can be overcome. Wave power is captured by buoys anchored to the seafloor offshore. Wave power has the classic problem of the difficulty of harnessing a very dilute energy source. Tidal power is generally captured in estuaries, using dams that allow the rising tide to raise the water level and then using turbines to create electricity as the tide goes out. Tidal power works well when you have a really good site, such as the Rance River on the French side of the English Channel, and pilot projects are starting to emerge in California.

Alternative forms of liquid fuel for powering vehicles (biofuels) are now being developed. Ethanol is an energy-rich liquid that can (theoretically) be produced from various plant products (sugarcane, corn, wood, agricultural waste, fruits). Biofuels are relatively clean-burning, renewable sources of conveniently transportable energy. They are carbon-emitting energy sources but are cleaner than coal. The technology is still being perfected, but the fact that most cars in Brazil run on an 85/15 blend of ethanol and gasoline is an indication that these renewable sources of energy may be a part of the transportation future.

There are several technical and social questions regarding biofuels: What plants will be the sources of energy? The easiest source to process, and thus the first to come onto the market, is sugarcane. One problem with using sugarcane to produce fuel is that it takes prime agricultural land to grow sugarcane, land that could be used to feed people. Will we one day face the choice between driving or feeding people? It takes energy to run the tractors and make the fertilizer to grow corn or sugarcane. Is the energy output more than the input? With sugarcane there is probably a net gain in energy. With corn, if you count the energy required to produce the fertilizer, then there is a net overall loss of usable energy. Without federal subsidies to farmers in the United States and ethanol import taxes, corn-based biofuels may not turn out to be profitable in the United States where farming costs are high.

The other important issue for sustainability is that growing biofuels competes with food production, contributing to higher global food prices. For about 100 years, people have tried to produce a biofuel that targets the cellulose in plant cells rather than the glucose. If "cellulosic" biofuels became practical, then entirely new sources of fuel would be available, using wood waste, paper pulp, and other current "waste products" to supply energy needs. Unfortunately, it is technically difficult.

In addition, there are infrastructure challenges in the transition to non-fossil-fuel energy. Batteries for energy storage, for example, raise concerns about recycling and disposal of the batteries and the environmental impacts and labor practices associated with mining the raw materials. There is no clean energy revolution without trade-offs and challenges.

All forms of energy that power human activities have advantages and disadvantages. How to evaluate these trade-offs is a complex, often technical, always political problem. Since the discussion must take place in the political arena, which can limit the depth and quality of the discussion, completely rational energy policy is seldom achieved. These policies, however, do influence how the state and country address global environmental issues that cannot be solved by one country alone.

GLOBAL ENVIRONMENTAL CHALLENGES

Climate Change

The global climate has a long and complex history of dramatic changes, with cold glacial cycles alternating with warm interglacials. Greenhouse gases, such as carbon dioxide, methane, and water vapor, naturally trap the sun's energy in our atmosphere, creating a "blanket" that keeps Earth warm enough to support life. This delicate balance has shifted over time, with greenhouse gas levels fluctuating during ice ages and warmer periods, playing a key role in maintaining a habitable planet. The North American continent has at times hosted tropical forests, been buried under sheets of ice nearly a mile thick, and been a desert so dry it supported virtually no plants or animals. However, the anthropogenic causes of modern climate change are acting on a much faster timescale than natural climate-changing events in the past.

Since the beginning of the Industrial Revolution, human production of carbon dioxide has increased dramatically, and the current disruption in climate is a result of the combustion of fossil fuels, which releases greenhouse gases, including carbon dioxide, methane, and nitrous oxide, into the atmosphere.

By examining gas concentrations, including CO_2, in bubbles within ice cores spanning vast geological timescales (up to 800,000 years as of 2021), scientists are able to reconstruct historical atmospheric conditions. They can then correlate CO_2 with past global temperature trends determined from other measures, such as oxygen isotope ratios, and ascertain the role of CO_2 in influencing global warming, aiding our understanding of climate change today.

Today, the observed increase in CO_2 concentrations in the atmosphere is produced largely by our removing fossil fuels from long-term storage in the ground and burning them to meet an annually increasing

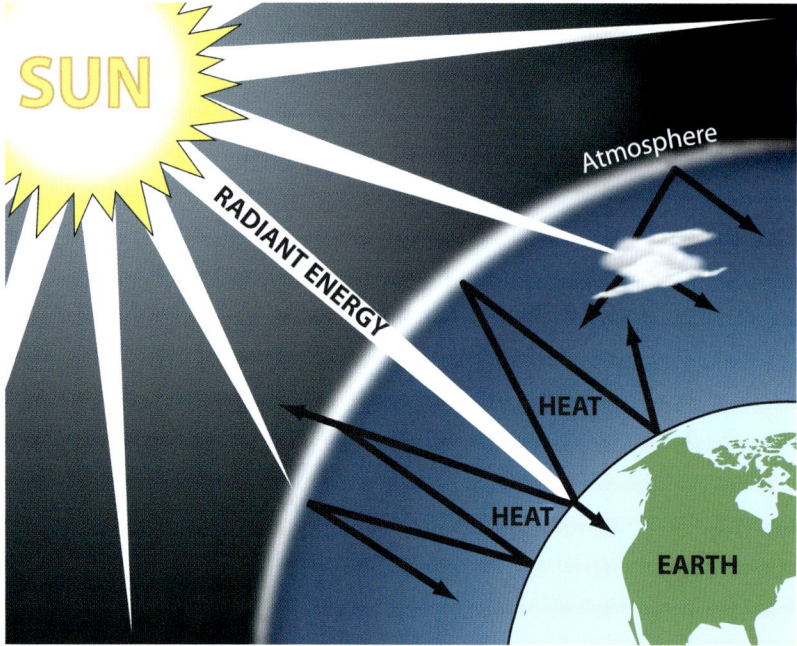

The greenhouse effect. The atmosphere protects organisms from solar radiation, and it insulates the planet from heat loss. Courtesy of the State of Washington Department of Energy.

demand for energy. The additional greenhouse gases are accumulating in the atmosphere to dangerously high concentrations.

These gases trap heat emitted from the Earth's surface. While carbon dioxide does not trap as much heat as methane and some other gases, it is still the most worrying greenhouse gas because it can remain in the atmosphere for 200 years, twice as long as nitrous oxide and almost 17 times longer than methane. For this reason, nitrous oxide and methane are sometimes referred to as short-lived climate pollutants. About 82 percent of the greenhouse gas currently emitted from the United States is carbon dioxide. By continually trapping heat, these gas molecules in the atmosphere regulate the temperature on the surface of the planet. As the concentration of greenhouse gases in the atmosphere increases, more heat is trapped and climate impacts intensify.

Continuing our current rate of CO_2 emissions could lead to a concentration of 500 parts per million of CO_2. This concentration of CO_2 could initiate a series of climate feedback mechanisms resulting in a

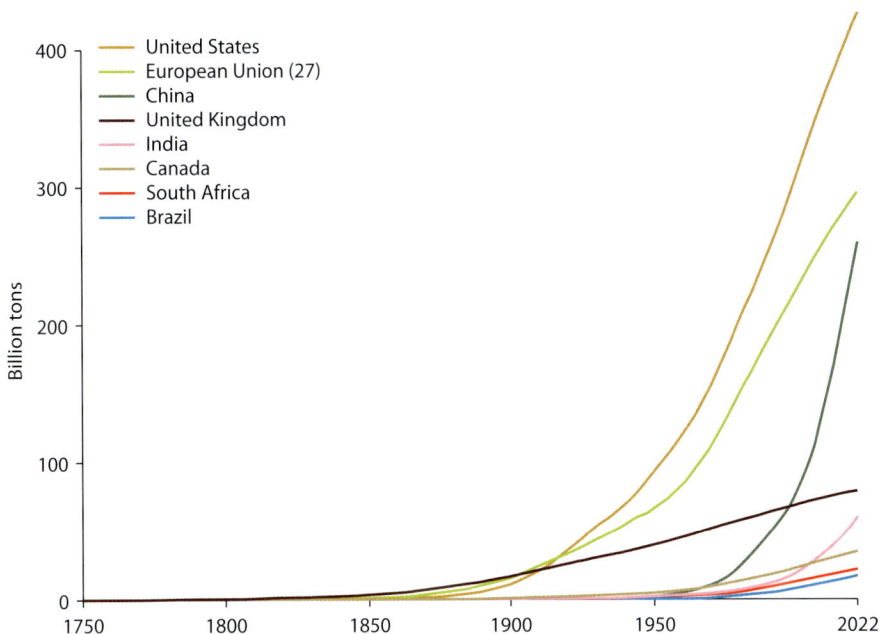

Total cumulative emissions of carbon dioxide (CO2), excluding land use change, since the first year of available data, measured in billion tons. Adapted from Global Carbon Budget (2023) with major processing by Our World in Data.

"hothouse Earth" scenario with temperatures of about 7.2 to 9 degrees (4 to 5 degrees C) higher than today and sea levels rising by 30 to 100 feet—sufficient to submerge many of the world's population centers. It is imperative that we lower the concentrations of these gases and allow more heat to escape to space to minimize these impacts.

While efforts in California and elsewhere are underway to limit emissions by moving to an all-electric economy powered by renewable energy, much more must be done. We must change our patterns of land use and consumption to reduce the overall demand for energy.

People living in highly developed countries have a far greater per capita impact on the atmosphere than people in less-developed countries. Hence California and other wealthy societies have a moral and ethical responsibility to reduce energy use and help others end their use of fossil fuels. A combination of expanding renewable energy, instituting regulations to reduce emissions, and changing behavior will all be necessary to achieve the very ambitious goals necessary to confront this existential threat to humanity, as well as our coinhabitants on the planet.

Climate Change in California

People around the world and in California are already experiencing warmer temperatures, hotter and more frequent heat waves, melting glaciers, rising sea level, and altered patterns of precipitation due to the human-induced warming of our climate.

In California, temperatures have risen by at least 0.9 degree throughout most of the state, surpassing 1.8 degrees in certain areas, including the low deserts where there has been a 4.5-degree increase. This ongoing warming, coupled with shifting precipitation patterns and reduced snowpack, poses significant challenges for native species, local communities, and agriculture.

Across California, changes are evident. The Sacramento River's peak runoff happens nearly a month earlier than it did in the early twentieth century. Glaciers in the Sierra Nevada have lost a staggering 70 percent of their area, reducing spring runoff—a vital source of fresh water. Grizzly Glacier, the last remaining glacier in the Klamath Mountains, broke apart in 2015 and is no longer considered a functional glacier. As a consequence of early snowmelt in the spring, soils are drying earlier and staying dry longer; as a result, crops require irrigation earlier and longer. Many parts of the state already feel water stressed, but with climate change and land use changes, dramatic declines in the availability of fresh water are expected. Decreases in snowpack are leading to reduced flows along the Colorado River. Already, Lake Mead has a water deficit of 1 million acre-feet per year, and with severe drought and continued overuse of Colorado River water, the lake levels could become too low for water to flow downstream.

Our ecosystems, both on land and in the ocean, are responding to climate disruption and experiencing local disappearances of species, forced alteration of geographic range, and new management challenges. Birds are changing their wintering patterns, moving farther north and closer to the coast. As climate change alters habitats, some species are being forced to shift their ranges to survive, and they may no longer persist in protected areas that were originally designated for their conservation. Native plants are blooming earlier, moving to cooler climes, and in some cases, losing their pollinators as ranges diverge. Some species may not be able to adapt locally to rapid climate change, and due to specific habitat requirements, such as soil type, they may also be unable to readily shift their range.

An increasing number of people are undergoing trauma associated with climate change, such as catastrophic fires, intense drought, flooding, accelerated erosion, and severe heat waves. Trauma-aware and informed approaches to discussing these extreme events are crucial, as these disasters can have profound psychological impacts on individuals and communities. Acknowledging and understanding the trauma associated with these impacts is important for everyone discussing climate change and interpreting its importance to others so that those affected can share their experiences and perspectives in a supportive environment. By integrating trauma-informed practices into these discussions, we can better address the experience many people are having as survivors, and foster resilience in the face of future challenges.

Grief is another emotion many people feel as they see beloved places change. One of the tools people use to cope with grief, sadness, and loss is to sing. We call songs in this genre laments. Here are the words to a lament some naturalists relate to:

THE WOOD THRUSH'S SONG BY LAURIE LEWIS

I walked down the hall where the woods used to stand
Concrete at my feet, brick walls at every hand
and over my head steel girders so strong
where I first felt the spell of the wood thrush's song

Now the wood thrush is vanished, seeking the place
that's not felt the crush of mans' embrace
the steep woods are gone now and oh how I long
to again feel the spell of the wood thrush's song

Over my head just a few years ago
the poplar leaves shivered when the breezes did blow
Now the deep hum of engines drowns the soft sigh
of the wind in the leaves of the few trees near by

Man is the inventor, the builder, the sage,
the writer and seeker of truth by the page,
but all of his knowledge can never explain
the deep mystery of the wood thrush refrain

Now the wood thrush is vanished, seeking the place
that's not felt the crush of mans' embrace
the steep woods are gone now and oh how I long
to again feel the spell of the wood thrush's song

Words and music: Laurie Lewis/Spruce and Maple Music, ASCAP

California's Future Climate

Current conditions in California are an early indication of the changes that are likely to come in the future. In mid-century California, a temperature increase of 4.3 degrees (2.4 degrees C) is anticipated under a low-emissions scenario, while a business-as-usual scenario predicts a rise of 5.8 degrees (3.2 degrees C). By 2100, the expected temperature increases are 5.6 degrees (3.1 degrees C) and 8.8 degrees (4.9 degrees C), respectively. These changes will result in warmer winters and more frequent and prolonged summer heat waves. Regarding precipitation, climate models exhibit uncertainty but show a tendency for more rainfall in the north compared with the south. It is crucial to recognize that year-to-year precipitation will become more unpredictable, dominated by extreme events such as droughts and atmospheric rivers, sometimes back to back, creating weather whiplash.

Heat wave days are projected to increase by 20 to 30 days per year in the Central Valley, and using water for heat protection may not be possible. These changes aren't just about the weather; they impact all aspects of California's society. Different regions may experience varying effects, but the overall picture includes direct impacts on public health and infrastructure due to rising temperatures and sea levels. Droughts, floods, and wildfires will continue affecting people's lives and economies. Tourism and the crops that drive California's economic success will face challenges due to changing weather patterns and more extreme conditions. A UC Berkeley study found that the economic damage to the state due to global temperature increases could have an annual price tag of $7.3 to $46 billion, and that $2.5 trillion of real estate assets will be at risk.

Other projected impacts in California include the following:

- Sea level rise, coastal flooding, and coastal erosion: Over the twenty-first century, sea level is expected to rise substantially. In combination with storm surges, sea level rise is expected to severely erode Southern California beaches.

- Higher risk of fires: Climate change makes forests more vulnerable to fires by increasing the length and intensity of the dry season. Hospitalizations from inhalation of wildfire smoke will rise.

- Damage to agriculture: More severe droughts will decrease soil moisture, further aggravating atmospheric warming and threatening California's agricultural industry, which generated nearly $60 billion in cash receipts in 2023.

Flames of the Simi Valley fire burn a Southern California mountainside. US Air Force photo by Senior Master Sgt. Dennis W. Goff.

- Increased demand for electricity: Higher temperatures and heat waves will drive up demand for cooling.

- Public health impacts: The number of days of extreme heat events will increase, causing more dehydration, heatstroke, heart attacks, and respiratory problems.

- Impacts to low-income and minority communities: Global warming's impacts disproportionately affect low-income and minority communities in California, who have the least ability to adapt to the impacts.

- Habitat modification, destruction, and loss of ecosystems: Climate change will adversely affect plant and wildlife habitat and the ability of the state's varied ecosystems to provide services important for human well-being. Prolonged drought has stressed trees especially in Southern California, leaving pines, firs, and hemlocks susceptible to bark beetles, resulting in high levels of mortality. In Yosemite National Park, pikas are being forced to higher elevations. Butterflies in the Central Valley have been arriving earlier in the spring over the past four decades. Warmer and more acidic ocean water will continue to alter marine ecosystems, disrupting food chains and coastal currents.

- Water shortages: The Sierra snowpack provides up to 65 percent of California's water supply. Sea level rise will cause saltwater intrusion into the Bay-Delta, threatening the water supply for 25 million Californians and thousands of acres of farmland.

These threats demonstrate the seriousness of the problem and the need to adopt policies to address and mitigate the sources and impacts of climate change. We are in a situation where every tenth of a degree matters. However, there is good news among the bad. California has led the nation in a commitment to reducing greenhouse gas emissions, working toward fossil-fuel-free electricity generation, as well as prioritizing nature-based solutions for carbon sequestration and climate abatement, such as 30×30, an initiative to protect and restore 30 percent of California lands by 2030.

Community-driven activities, including environmental stewardship, participatory science, civic action, communication, and education, have the potential to foster innovative practices and cultivate a shared vision that leads to collective impact. Californians are acknowledging the unsustainable nature of the current situation and are embracing new approaches,

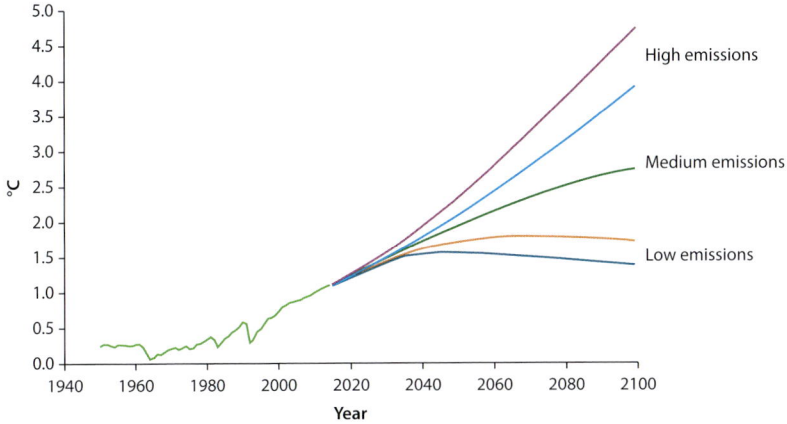

Historical mean global surface temperatures (1950–2014) and modeled projections starting in 2015 under low, medium, and high greenhouse gas emissions scenarios. Adapted from data for Figure SPM.8, *Summary for Policymakers of the Working Group I Contribution to the IPCC Sixth Assessment Report* (v20210809).

ranging from ecological restoration to the promotion of regenerative economies focused on thriving rather than pursuing infinite growth.

California is often a global trendsetter. Californians are working on ways to adapt to these changes, and together we can create new economic models and lifeways. As naturalists, we have the opportunity to deepen our connection with nature, strengthen our interdependence, and benefit the entire community of life.

The state's annual climate change assessments are the best way to keep up with our understanding of climate science and the threats we face in California. The UC Climate Stewards Program provides an opportunity for a deeper dive into climate change science and ways to advance collective action.

Dead Zones, Fertilizers, and Manure Management

Dead zones are low-oxygen areas in the world's oceans and lakes. Oceanographers first began noting dead zones in the 1970s. Dead zones, also called hypoxic zones, are caused by a process called eutrophication in which excess waterborne nutrients stimulate excessive plant growth or algal blooms, eventually depleting the oxygen in the water, creating biological deserts.

Collective Climate Action

Too often, we feel like the task of improving the climate falls to us as individuals, and while our habits do matter, it is through collective action that we can have the biggest impact. California communities are taking action to advance resilience, to help ecosystems and communities survive disruption, and to anticipate, adapt to, and flourish in the face of climate change.

Civic Engagement

- Share information about climate change and solutions with friends and family. Show up at public meetings. Join your neighbors in action. Ideas for communicating constructively and effectively are found in chapter 8.
- Help implement a Community Choice Aggregation (CCA) energy program so residents can purchase clean energy and help make it affordable for all.
- Encourage cities and counties to insist on and incentivize solar design on all new construction.
- Support bans on gasoline-powered leaf blowers and other polluting machinery.
- Urge your city to add bicycle lanes and walkways and make them safe. If you are a parent of school-age children, start a walk-to-school group.
- Advocate for policies that avoid the use of fossil fuels, single-use plastics, and other contaminants.
- If you own stock, join shareholder actions to petition for climate accountability from the company.

Land Stewardship

- Start or join a prescribed burn association to put good fire on the land.
- Advance LandBack opportunities for Tribal sovereignty and Indigenous stewardship.
- Control invasive species.
- Restore native plant and animal populations.
- Recover wetlands by restoring natural streamflow.
- Protect and restore wildlife corridors to facilitate the movement of organisms and range shifts under climate change.
- Provide climate refugia by planting and caring for native plants. Create and tend urban green spaces to provide habitat for native species.

(continued)

Collective Climate Action *(continued)*

- Reduce urban heat islands by installing cool roofs and streets.
- Encourage efforts to capture stormwater, reuse wastewater, and recharge groundwater.

Food, Agriculture, and Waste

- Establish and tend urban gardens.
- Reduce food waste from farm to market to table.
- Join or start a gleaners group.
- Participate in beach, road, and other cleanups.

Hypoxic zones can occur naturally, but the rapid rise in dead zones in recent decades is the direct result of human activity, primarily driven by agricultural and industrial activity such as fertilizer, manure, and sewage pollution. The process is exacerbated by rising global temperatures.

There are hundreds of dead zones in the world's oceans. They appear periodically off the coast of Southern California, and overall dissolved oxygen levels have been declining there since 1984, causing concerns about the impacts on marine organisms and habitat. In the case of California hypoxic events, the causes are a complex interplay of warmer ocean temperatures, currents, upwelling, and nutrient runoff.

Fertilizers, improperly treated manure, and sewage are important sources of water pollution and contributors to harmful algal blooms and hypoxic zones. Manure management will be an increasingly important problem on spaceship Earth. In addition, animal waste releases methane gas. Clearly the day has arrived when we can no longer afford to regard our atmosphere and our rivers as global dumping grounds.

Agricultural Issues

California is the nation's agricultural powerhouse. California's farms produce nearly half the vegetables, fruit, and nuts grown in the United States. For some crops—such as almonds, artichokes, table grapes, pistachios, olives, walnuts, and figs—California grows 99 percent of what is commercially produced in the United States. California is also the

The Sierra Nevada in a Changing Climate

The majestic Sierra Nevada is revealing changes associated with climate change that are impacting its unique ecosystems. Long-term warming trends, dwindling snowpack, and alterations in streamflow are painting a new reality for these high-elevation regions.

Over the past century, the Sierra Nevada has warmed by an average of 1.8 to 3.6 degrees (1 to 2 degrees C), with Yosemite experiencing an alarming 5.4 degrees (3 degrees C) increase. The impact on wildlife is unmistakable. Birds, in particular, are on the move, seeking refuge in cooler climates. The Savannah sparrow has one of the largest documented range shifts among California birds, with its upper elevation limit moving 8,211 feet upslope from any limit observed prior to recent climate warming. Other birds, including the red-winged blackbird and western meadowlark, have moved their ranges downslope.

These movements signify adaptation, a positive sign from a conservation perspective. Yet, questions arise about the species that remain static. How are they coping with changing climate? Early research suggests that some are adjusting by altering their reproductive cycles, laying eggs and hatching chicks earlier in the year to cope with changing weather patterns.

Looking ahead, projections indicate a continued warming trend of 5.9 to 9.0 degrees (3.3 to 5.0 degrees C) in the Sierra Nevada by the

(continued)

nation's largest dairy producer and supplies vast amounts of other farm products, including greenhouse and nursery products and cattle. Simply put, California is the leading agricultural producer in the United States in terms of cash farm receipts.

Of the state's 100 million acres, 24 million acres were dedicated to some form of agriculture in 2022, and its Mediterranean climate, freshwater resources, and soils allow for higher intensity of production compared with other parts of the country.

Like so many concerns in California, agricultural land issues and pressures reflect the complexity of the state as a whole. Population growth, land use, immigration, worker safety, water wars, nonpoint source water pollution, carbon credits, and many other disparate issues find their nexus in California agriculture. One of the most pressing concerns for food production is the development and conversion of California farmland for suburban and urban uses. As the state's population

The Sierra Nevada in a Changing Climate *(continued)*

end of the twenty-first century. This anticipated rise in temperature could elevate the rain-to-snow transition line by 1,500 to 3,000 feet, making snowfall below 6,000 feet increasingly unlikely. The consequences are significant—diminished snowpack, altered streamflows, and drier soils with 15 to 40 percent less moisture than historical averages. Downhill skiing could become a memory!

The aftermath of the severe drought from 2012 to 2015 serves as a stark reminder of the vulnerability of California's high-elevation ecosystems. Pines, firs, and hemlock trees suffered under the prolonged stress, rendering them susceptible to bark beetles. The ensuing beetle infestation claimed an estimated 147 million trees, particularly in Southern California, where drought stress was most severe.

The loss of these trees not only affects the forest's ability to sequester carbon, but also increases carbon emissions as dead trees decompose. Furthermore, the abundance of dead trees becomes a fuel source for wildfires, increasing risks to structures and human safety. Balancing the need for conservation with the necessity of maintaining habitat for cavity-nesting birds and other wildlife presents a complex challenge.

Conservation efforts, sustainable practices, and informed decision-making are essential to mitigate the impacts of climate change on California's high elevations.

expands and land prices rise, pressure to convert farmland to housing and commercial uses grows.

Since many cities in California are located adjacent to prime cropland, urban expansion, particularly exurban development, directly reduces the farmland base. According to the California Department of Conservation, between 2016 and 2018, California lost 56,186 net acres of farmland to other uses. The majority of what was lost was "prime" agricultural land, and California could lose nearly 5 million acres of farmland if conversion to urban uses continues at this rate.

Climate change in California is likely to have significant impacts for agriculture, including a less reliable water supply, a decline in the cool nights necessary for some crops, and a shift in where certain crops can be grown. For example, according to research done at UC Davis, some areas, such as Yolo County, will become too hot over the next 50 years to grow crops that many farmers rely on, such as tomatoes and cucumbers. In addition, the reduction in the number of cool winter nights with temperatures below 44.6 degrees (winter "chill hours") can lower yields on tree fruits and nuts by disrupting pollination and flowering.

Agricultural Water Use

As noted in chapter 3, agriculture as a sector uses 80 percent of developed water sources in California. With never enough to go around, conflicts between urban, environmental, and agricultural water needs are inevitable and have been evident in California for decades. California's water delivery system has allowed farms to flourish in areas that would not be able to support agriculture if the farms had to rely on local surface and groundwater sources.

Water use by agriculture can also sometimes trigger ecological concerns. For example, in dry years, excessive pumping from wells and springs can collapse aquifers or cause saltwater intrusion. In addition, ground and surface water can be contaminated by surface runoff of fertilizers and pesticides, polluting the water. Construction and maintenance of buffer zones of native vegetation next to water sources can help to mitigate this problem.

Agricultural water use efficiency has risen substantially in recent years to address reductions in water availability during droughts and to incorporate practices necessitated by local Groundwater Sustainability Plans. Agricultural water use efficiency methods include conversion to less water-intensive crops or dry farming, drip irrigation, sensor-based automated irrigation scheduling, soil management practices such as no-till, compost applications, and cover crops that increase soil's water-holding capacity. While much of California agriculture already uses some or many of these techniques, a study by the Pacific Institute has indicated that by maximizing water use efficiency methods, "California's agricultural water use could be reduced by 17–22%, while maintaining productivity and total acreage irrigated."

Sustainable Agriculture and Carbon Sequestration

How agriculture is practiced can make a big difference in its impact. Practices associated with monoculture, including the use of pesticides and herbicides, can take a toll on biodiversity as well as impact the health and well-being of farmworkers and communities.

Sustainable agricultural practices can benefit both the environment and a farm's longevity in the face of climate change. Sustainable agricultural practices include

- Mulching
- Compost application
- Native plant hedgerows
- Riparian vegetation buffer zones
- Cover cropping
- Crop diversity and crop rotation
- Conservation tillage, including reduced and no tillage
- Dry farming
- Agroforestry
- Whole orchard recycling

In addition, a focus on the three pillars of sustainability goes beyond environmental sustainability to include social sustainability for workers and communities as well as economic sustainability connected to the maintenance of small and medium-sized farms.

The use of specific on-farm practices in order to take carbon out of the air and store it in soils and plant material is sometimes called carbon farming. Carbon farming uses some of the methods mentioned above with the goal of maximizing soil health and sequestering carbon, while also increasing soil water-holding capacity and encouraging biodiversity by attracting predators, pollinators, and other organisms.

The potential is huge: California's vast agricultural acreage represents an enormous opportunity for climate resilience, groundwater recharge, wildlife habitat enhancement, and carbon sequestration.

Opportunities for using soil and plants as carbon sinks go beyond traditional agriculture. According to the USDA California Climate Hub, "it is estimated that over 25 million metric tons of CO_2 can be sequestered annually on natural and working lands in California by 2045." A 2018 nationwide study found that land management tools such as reforestation, improved forest management, protection of forests and grasslands from conversion to other uses, habitat restoration, and other conservation practices have the potential to sequester up to 1.2 billion metric tons of CO_2 annually.

Air Quality

There are several types of air pollutants of interest to naturalists in California, including smog, particulate matter, sulfur dioxide, carbon monoxide, and carbon dioxide.

Smog in western states is mostly ground-level ozone, formed from nitrogen oxides (NO_x) and volatile organic compounds (VOCs) in the presence of sunlight. NO_x comes from combustion sources such as vehicles and power plants and from soil fertilization. VOCs come from such sources as evaporating fuel, vegetation, and consumer products. Ozone is worst on warm, stagnant days. Although originally confined to urban basins such as Los Angeles, ground-level ozone is now distributed regionally and is transported between continents. For example, rural areas in far northern California receive some ozone from plumes that travel across the ocean from Asia and make landfall. Ozone causes difficulty in breathing, worsens asthma, and over long periods, reduces lung volume. It also damages plants by reducing photosynthesis, growth, and root development. Many plants exhibit a bronze speckling on the upper leaf surface when exposed to ozone.

Particulate matter can consist of many chemical constituents that come from diverse sources, including wood burning, diesel engine emissions, road and agricultural dust, and most notoriously, large wildfires. Soot in smoke is a fine particle that may be carcinogenic and degrades visibility. Fine particles of less than 2.5 micrometers (also known as PM2.5), like the kind found in wildfire smoke, are of most concern to health. They have been linked to premature death, particularly in vulnerable populations of the very old, the very young, and those with poor health, primarily through worsening cardiovascular symptoms. In general, particulate matter is not damaging to vegetation, though when it contains nitrogen, it may serve as a fertilizer (even leading to harmful nitrogen saturation of ecosystems) when it deposits on vegetation. Exceptions are cement dust, which is alkaline and corrosive to plants, and materials like heavy metals (lead, copper, mercury), which when present can be harmful to grazing animals and consumers of milk and meat. Most particulate matter is in the form of fine particles.

Sulfur dioxide (SO_2) comes from burning fuels that contain sulfur, such as in coal-fired power plants. SO_2 is very irritating to the airways, may trigger asthma attacks, and when oxidized to sulfuric acid, contributes to acid precipitation. Acid precipitation may acidify lakes, particularly in granite landscapes such as the Sierra Nevada where natural pH buffering is weak.

Addressing Global Challenges

While pollution and climate change can be demoralizing, humanity has provided examples of effective global action. In the 1980s, a hole in the ozone layer was determined to be expanding due to the use of chlorofluorocarbons (CFCs). CFC production was banned in 1990 by signatories to the Montreal Protocol, and this global response has been effective at dramatically reducing CFC use and impact. Other compounds, including those with bromine, have been found to have the same ozone-destroying effect in the stratosphere, and current efforts focus on ODS—ozone-depleting substances. The worldwide reaction to solving this problem may be the first example of a global political response to an environmental threat. It is an example that can give us courage and confidence in confronting the even larger problem of climate change.

Carbon monoxide is a by-product of incomplete combustion, whether in automobile engines or from campfires or other biomass burning. It may accumulate, for example, in valleys with little air movement, in caves, and in enclosed tents or automobiles. It causes headache and chest pains and can lead to rapid death in humans but surprisingly is not harmful to plants.

Carbon dioxide (CO_2) is also considered an air pollutant. It comes from combustion, in California mostly from vehicles, as well as from many natural processes, including animal respiration.

Warmer temperatures due to climate change will lead to worse ozone air pollution in many places, because more VOCs are emitted at warmer temperatures from plants and evaporating fuels and because the reactions that form smog increase with temperature. More visibly, increases in the size and intensity of wildfires and their resultant plumes of smoke have memorably spread not just downwind of the fire but across the state and the nation, causing apocalyptic-looking red skies in the most extreme cases.

Solid Waste

In nature, there is no such thing as waste. A tree dies and as it decomposes, it becomes food for fungi, a home for insects and birds, and later, a seed log for new trees. In human society, however, people treat

Students work to maintain a compost bin. Photo used by permission. © Regents of the University of California.

materials as if they can "go away," and we excel at producing waste. The average American generates approximately 35 pounds of municipal solid waste per week, which translates to about 300 million tons for the whole country annually. But what happens to that waste? How much of it is or could have been recycled, reused, or composted? And where did it come from in the first place?

Every item you buy has "embodied energy." Embodied energy is the energy that was used to get the product from its source to the consumer: to extract the resources the product is made from (wood from logged trees, plastic from refined petroleum); to manufacture, package, and advertise the product; to ship it from the manufacturer to the store and then to the home; and finally, to collect it for recycling or landfilling. For every product in your house, a certain number of gallons of gas and water were used to create, transport, and dispose of it. When items are thrown away, what is wasted is not only the materials composing the product, but the resources that went into creating it.

Municipal solid waste typically goes to a transfer station or materials recovery facility where it is sorted and recyclable items are removed from the waste stream. The rest is trucked to a landfill. A landfill is a gigantic hole that is lined with clay or plastic to prevent leakage into the soil and groundwater. Landfills are an inevitable part of an economy where manufactured goods are disposable.

Material Waste and Material Cycles

How would you reimagine the solid waste problem from a nutrient cycles perspective? We are removing materials (minerals, wood, metal, oil) from the planet's reservoirs at an accelerated rate, using some portion of them and then mixing them together for long-term storage in a landfill. Along the way, during removal, manufacturing, use, and disposal, parts of the materials (pollutants) are discharged into the air and water to be reabsorbed into the system at the natural rate. What cannot be readily reabsorbed is left to accumulate in the atmosphere or taken up in excess by living organisms, contaminating and sometimes killing them.

Landfills pose numerous problems. First, they are built on land that could be used for other purposes. As locations for new landfills become scarce, building new ones is harder: few people want a landfill located near their house, and increasingly, the land is desired for other purposes such as housing and open space. For this reason, many cities truck their garbage far away at great expense.

Second, landfills also generate something called leachate, which is a toxic liquid that forms when rain or groundwater gets into a landfill and leaches out chemicals from the waste and when liquid is produced as the waste settles and decomposes. Although landfills have liners to prevent the leachate from contaminating groundwater, those liners are not perfect and break down over time. To minimize the possibility of moisture getting in, the materials in landfills are covered to shield them from rain, wind, sun, and air. The result is that many things in a landfill take decades, if not hundreds or even thousands of years, to decompose. A plastic bag may take between 500 and 1,000 years to completely break down.

Third, as the materials do degrade, landfills emit methane, primarily from food waste, yard debris, and other organics. Although California landfills are set up to recover methane and use it for productive purposes, according to CalRecycle, landfills are the third-largest source of methane in California. Organics like food scraps, yard trimmings, paper, and cardboard make up half of what Californians dump in landfills, and organic waste in landfills emits 20 percent of the state's methane.

In 2022, the state generated 76 million tons of solid waste. Of that, nearly 41 percent was recovered in some manner (source reduced,

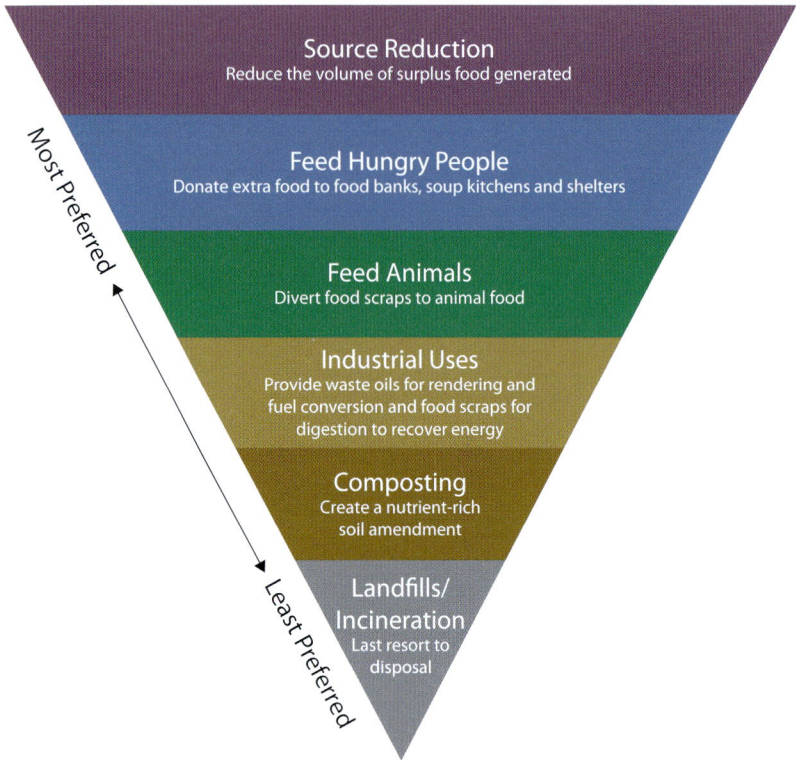

Food recovery hierarchy. Note that composting is nearly at the bottom of the hierarchy, with source reduction and redirecting unused food for human and animal consumption as preferable actions. Courtesy of the California Department of Food and Agriculture.

recycled, composted, and other less common forms of recovery). Due to diminishing landfill space and in order to reach the state's climate goals, California has instituted laws that set the nation's most ambitious waste recovery goals. California has a goal of 75 percent recovery, a goal that was supposed to have been met in 2020, so there is still quite a bit of progress to be made!

Two laws stand out. SB 1383 institutes the state's short-lived climate pollutants strategy and establishes targets to reduce the volume of organic wastes destined for landfills. In addition, California must cut its methane emissions 40 percent from 2013 levels by 2030. Interestingly, the law specifically addresses disposal of edible food in order to redirect it for food reuse and sets up a food recovery hierarchy similar to the better-known recycling hierarchy (Reduce, Reuse, Recycle). According to CalRecycle, "3 million cars worth of climate pollution will be cut by reaching [the law's] recycling and food rescue targets."

Community efforts like Food Finders in the Los Angeles area are representative of one model for redistributing food waste to food-insecure individuals. They collect usable food cast offs from area restaurants, supermarkets, schools, and hospitals and distribute them to homeless shelters, food banks, and other organizations to feed people.

A second law of note is the Plastic Pollution Prevention and Packaging Producer Responsibility Act. According to CalRecycle, the law sets the first specific source reduction goals in US history. The goals set by the law are to reduce single-use plastic by 25 percent, recycle 65 percent of single-use plastic, and ensure 100 percent of single-use packaging and plastic food ware is recyclable or compostable by 2032. Most notably, the law shifts responsibility for reaching these goals from consumers to producers.

Movement toward meeting ambitious solid waste goals has been uneven. However, progress has been made. In 2003, San Francisco set a goal of achieving zero waste and has since cut its landfill disposal in half. Los Angeles County diverts 65 percent of its municipal solid waste from landfills.

Does it really matter? Yes. Reuse and recycling are not just feel-good activities. These actions conserve resources and directly reduce carbon emissions and help in the fight against global climate change. The EPA states, "Harvesting, extracting, and processing the raw materials used to manufacture new products is an energy-intensive activity. Reducing or nearly eliminating the need for these processes, therefore, achieves huge savings in energy. Recycling aluminum cans, for example, saves 95 percent of the energy required to make the same amount of aluminum from its virgin source, bauxite. The amount of energy saved differs by material, but almost all recycling processes achieve significant energy savings compared to production using virgin materials."

Approximately 33 billion pounds of plastic enter the oceans every year, and studies show that much of that plastic comes from land sources, essentially litter that makes its way to stream systems and then to the ocean. All over the world's oceans are huge floating patches of trash. The plastic breaks down into tiny pieces, making it hard to study or clean up but easy for sea life to mistake for food. The patches have a greater mass than all of the plankton available in the northern Pacific Ocean and are now a permanent and very damaging part of the aquatic food chain.

Easy, Free, No Equipment Required!

Some conservation efforts require political will, inputs of capital, specialized machinery, or long periods of time to be effective. Other forms of conservation are simple, are instantaneous, require no money, and can be done immediately. Every day, you can help make a difference. Turning off your tap, turning down your thermostat, walking to the store, buying less, using the food you have, picking up litter—these are small and meaningful acts.

Environmental Justice and Equity

Historically, poor and minority communities have borne the brunt of pollution, sea level rise, and discriminatory urban planning policies. Environmental justice acknowledges historical injustices related to redlining, land dispossession, and the resultant disproportionate exposure to environmental risks and inequitable access to resources such as clean air, water, and open space. Today, many underserved urban communities have very little access to natural areas. According to the LA County Department of Parks and Recreation, Los Angeles has half the median amount of parks and green space found in other high-density US cities, with most of the census blocks in the city having less than 3 acres per 1,000 people. Actions to address these injustices include policy reforms to protect communities from environmental factors that negatively impact their lives, as well as initiatives to promote equitable access to environmental resources and to mitigate the disproportionate impacts of pollution and climate change on marginalized communities. Having access to natural areas close to home is vital for human health and well-being. From small parks to urban forests, natural areas offer opportunities for nature observation, recreation, and conservation. These green spaces also help filter pollutants and provide shade and cooling as temperatures rise. Overall, prioritizing the creation and maintenance of natural areas is crucial, especially for underserved communities in cities.

There are efforts to change these dynamics, efforts coming directly from the affected communities. A wonderful example is the LandBack and Indigenous stewardship movement. LandBack is a growing movement focused on returning land to Indigenous Peoples, encompassing various actions to enhance their access to and stewardship of ancestral homelands. The movement seeks to address the long history of land dispossession and barriers to land access that came with colonization and the adverse effects on cultural vitality and intergenerational knowledge transfer within Native communities. LandBack allows Tribes and Tribal communities to steward the land in a manner that restores their reciprocal relationships to a place.

Through Governor Newsom's Executive Order N-15–19 (2019), California recognizes and apologizes for state violence against Native Nations. Since the executive order, California has provided conservation funding for Tribes and Tribal communities to advance LandBack. LandBack provides avenues for Indigenous Peoples to reclaim taken land, facilitating the restoration of healthy reciprocal relationships. The movement is gaining momentum to address justice, promote Indigenous stewardship, protect the community of life, and advance climate resilience. Actions taken under LandBack range from the full return of land title, without usage restrictions, to limited agreements for access. Significant funds are often required for Tribes and Tribal communities to secure land title today.

LandBack engagement is open to various entities, including individuals, religious institutions, land trusts, and local, state, and national governments. Recent State of California initiatives focusing on biodiversity conservation and climate resilience acknowledge the intrinsic relationship between Indigenous cultures and the land, highlighting the crucial role of Tribal engagement. The California Climate Adaptation Strategy underscores Tribal nations as critical leaders in land stewardship, habitat restoration, and natural resource management, making LandBack a hopeful climate action.

Many historically marginalized groups have for decades taken action to redress environmental inequities in their communities, efforts that too often are not widely recognized. One notable example is the West Oakland Environmental Indicators Project, a resident-led environmental justice organization using collaborative problem-solving to work with regulators and scientists. They use participatory research to collect data and develop strategies for air pollution reduction and community equity in one of the Bay Area's most environmentally impacted areas.

Population

So much about how the planet fares in the future is predicated on the size of the human population, the amount of resources we each consume, and how we consume them. According to the United Nations, population growth is expected to peak at 10 billion in the 2080s. How we address the many needs of all the people on the planet, fairly and equitably, is the task before humanity.

One of the more remarkable things about California is that even as the population increased, resource consumption decreased or remained the same for many measures, including water use and energy use. Collective action, policy changes, new technologies, community organizing, and visionary laws have helped reduce the per capita impact Californians are having on the environment.

The wonderful and terrifying thing about living on Earth in the twenty-first century is that we have powerful tools that make it abundantly clear that we live on one planet, with nowhere else to go. Virtually everyone has now seen images of the Earth from space, showing a blue sphere rotating in a huge, black void. We are rapidly reaching the limits of human population and resource use that the planet can support. We are creating global environmental dangers that no previous generation has seen or been challenged to solve. We also have unprecedented tools to perceive, communicate about, and potentially overcome these problems that no previous generation has possessed. Together, we can make them count.

Explore!

VISIT A POWER PLANT AND FIND OUT WHERE YOUR POWER COMES FROM

Call your local power company, and ask where your power comes from. Your power is likely to come from multiple sources, so find out if any of the power plants are local and if you can go see them.

VISIT YOUR LOCAL LANDFILL OR TRANSFER STATION AND FIND OUT WHERE YOUR TRASH GOES

Call your waste collector or visit their website to find out where your trash, recycling, and compost go. Be sure to ask what sorting happens on site and what goes elsewhere. For a landfill, see if you can find out how old it is and how full it is. Landfills are large operations and may require permission to visit. However, the operator of the site may have a public information officer who can help you.

(continued)

VISIT A LEED-CERTIFIED BUILDING OR TAKE A GREEN HOME TOUR

You can find a list of projects certified by Leadership in Energy and Environmental Design (LEED) at the US Green Building Council website's LEED project directory.

CALCULATE YOUR CARBON FOOTPRINT

Carbon footprint calculators are not perfect, but they can be thought-provoking and give a snapshot of ways to reduce your carbon output. The US EPA and The Nature Conservancy carbon footprint calculators online are good places to start.

DO A DUMPSTER DIVE!

How does your household garbage stack up? Write down everything you think you would find in your garbage can over the course of a week. Be specific and try to assign percentages, such as paper 10 percent, plastic 20 percent. Then actually go through your garbage the day before it's picked up, and sort it by material. (Use gloves and goggles.) Ask yourself, How much of this could have been recycled? How much composted? Why did I buy this product in the first place? Did I use it fully or even need it? How many of these items were disposable and in use for less than one day? One hour? Could I have bought something similar that could have been reused, recycled, or composted? In the following week, see what you can do to cut the waste stream by 50 percent, and see if you can maintain the lower production of waste for good.

LEARN ABOUT YOUR FOOD

The carbon impact of food is not restricted to the land on which it is grown. Growing food sustainably may sequester carbon, but processing it and shipping it emits carbon. Do some research on your food. Go to your kitchen and make a list of the countries your food items come from. For packaged foods, look for small print near the ingredients list for "Product of X." The international origin of many unlikely foods—crackers, cereal, canned goods—may surprise you.

8

Interpretation, Collaboration, and Participatory Science

Translating the language of nature into human meaning is the way naturalists share their sense of wonder in the natural world with others. Communicating knowledge is a key step in this endeavor: people use narrative and stories to convey relationships between organisms and the planet, how those relationships change over time, and how to work toward a positive future. For naturalists, this often takes the form of interpreting nature and engaging with others in environmental stewardship.

People have a strong interest in places they know. Placing the resource and land in context, acknowledging the land's original inhabitants, taking people for a hike, bringing an animal into a classroom, or giving a short talk at a visitor center are some of the ways naturalists share their passion for nature and place. Naturalists are also called upon to answer questions about the natural world in public settings. Interpreting scientific concepts and observations of the natural world for diverse groups are important skills for naturalists to develop. Sharing is also what engages many naturalists and motivates them to want to learn more.

Interpretation can be conducted in many different ways, from formal presentations to storytelling to shared cultural activities. Western scientific methodology has one method for teaching about the natural world and our place in it, while Native American Tribal traditions have another. The approach presented here is one set of ways, but there are many others.

Naturalists also possess knowledge that's useful to planners and policymakers. Learning to communicate with decision-makers and collaborate with a variety of interest groups ensures that nature conservation and ecological knowledge are considered in land use planning and other political processes that seek public input. In addition, naturalists possess skills that can be applied as part of participatory science projects. By interpreting and contributing to research, we advance our own understanding and encourage others to form a lifelong appreciation for nature.

Community-driven initiatives, including interpreting nature for others, civic engagement, collaborative conservation, and environmental stewardship, can result in collective impact for the benefit of biodiversity conservation, climate resilience, and human well-being.

INTERPRETATION: WHY, WHAT, AND HOW

People who do interpretation are highly motivated to connect with others and to protect the natural world. Encouraging enjoyment of nature, improving their own and others' health and well-being, and building community are among the motives for interpreting nature. In addition, interpretation allows organizations to express and meet their goals. For example, the National Park Service trains park interpreters, with the idea that they will help people care about National Parks. In Yosemite, for instance, some interpreters present programs on bears to help the public become aware of the animals' needs and behaviors and make sure that people store their food properly so bears stay wild, while cultural interpreters share some of the lifeways of Tribal communities and explain the ongoing connection to their homelands.

Interpreting nature goes beyond communicating scientific content. The larger goal is to convey a general theme or issue by sharing information in a meaningful and pleasant way that is relevant to the audience. The core objective in interpretation is always connecting the audience to a bigger story, rather than just dispensing information, so interpretation should be about forming connections rather than presenting facts. The best naturalists guide people to the brink of discovery and then get out of their way and let them immerse themselves in their own experiences. Truly engaging people by facilitating an experience they consider worthwhile leads them to want to experience more.

In thinking about interpretation, we need to keep in mind three important things. First, multitasking can be self-defeating. For example, listening to music on the trail will distract us from the sound of an eagle protecting its nest. Second, all knowledge is socially constructed. Much of what we learn is by direct transmission from others. We interpret,

A group of adults and children play a game to learn about a vernal pool in Sonoma County. Photo courtesy of Greg Damron.

refine, validate, and reinforce our experiences by describing them to other people, and by reexamining them. Third, everything is learned by connecting it to something we already know. It's by making connections with prior knowledge that we fit new learning into our scheme of the world. Each of us holds a mental map of the world in our brain. As we grow in experience, our mental maps become more nuanced, detailed, complex, and refined. It then becomes easier for us to put new information into context and to gain understanding. This is in part why adults put a premium on information that falls in line with their worldview, values, and existing knowledge, and people of all ages lend more attention to trusted sources of information, especially if they share values and aspects of their identity. Hence the importance of representation by a diversity of people when extending information to the public.

To deepen observers' understanding, interpreters can facilitate direct experiences like sitting quietly, watching carefully, and listening. Many parks and nature reserves have observation platforms for exactly this purpose. Interpreters often allow people to cultivate their own experience, and only later ask or answer questions. The hope is that people will connect with the experience, not just hear information.

In all cases, it is important for an interpreter to convey respect for the land and for the audience, to convey that we are guests in this place, and to frame the idea that the local environment has nurtured people for tens of thousands of years and plants and animals for millennia.

Interpretation and the presentation of information is related to teaching, and some teachers are great interpreters. An example was the late Professor Ken Norris from UC Santa Cruz, who created a program in which college students traveled around California on a bus with their professors, stopping to observe the natural world in diverse habitats around the state and to ask questions. Norris rarely lectured or told his students "how it is." He preferred to set up projects that allowed his students to have their own experiences, develop their own questions, and seek their own answers. Norris asked his students to keep journals, and he loved to read their musings aloud. He assigned students to spend 4 hours sitting alone in one place, just observing. Another of his "assignments" was to have students choose a plant, any plant, and spend an entire day watching it and thinking about that plant. They might ask, Where does that plant grow? In the sun, or in the shade? On rocks, or in deep rich soil along a creek? Norris used to say to his students, "The object is the authority." In other words, make time to experience a bird, plant, or rock before hearing what an expert has to say about it.

Of course, personal experience needs to be balanced with reading and communicating about organisms and their environment. Reading, writing, and videotaping can be powerful tools for social transmission of knowledge, allowing us to learn from the experiences of people from other cultures, places, and times. Imagine speaking with someone who has interacted with an extinct species or seen a river that has changed its course in their lifetime!

One of the best ways to learn about the relationship of the present to the past is through oral history and story. Learning and communicating are not just intellectual exercises but occur in a reciprocal relationship between naturalists and their peers or an audience. Each time a story is told, it breathes life into the content knowledge a naturalist is attempting to share. Narratives are also much easier for people to comprehend, in part because they engage the listener and are interesting to listen to. There is strong evidence that people can digest more material and recall it better, regardless of how familiar or interested they are in the topic, if it's conveyed in a narrative format.

Culture and traditional ecological wisdom are also shared through story, usually expressed as a narrative and through shared experiences, which is a way of showing, not telling, others what needs to be done. In other words, people are more likely to start or continue an activity if we show them through example what can be done and the benefits of these actions, rather than telling them what to do. Anyone who has raised children can attest to the fact that a child may resist being told what to do, but if they see other children in action, they are likely to jump in.

Naturalists typically don't have captive audiences. Their groups often are on vacation or enjoying free time, want to be entertained and have fun, and may not have strong motivation to "learn." Participants can and will leave, physically or mentally, if they are not engaged. To keep people engaged, interpretive presentations need to be relevant, meaningful, and enjoyable. People learn best when they're comfortable. Humor, used well, can help make an audience comfortable and also engaged. Another challenge interpreters face is that questions asked and issues discussed can come from many different disciplines. So, it's important to remember that interpreters can offer a pleasant, informative, and meaningful experience without knowing all the answers. It's fine to simply note that those are not details you're familiar with, and perhaps suggest another source of information for follow-up. Perhaps the most important sentence interpreters need to be able to speak is "I don't know."

Ask any group of seasoned interpreters how they create presentations, and while their wording will vary, they'll usually express a core list of approaches:

- Get to know the place or thing the program will focus on: Hike, read, talk with others, and follow one's passion and the organization's mission in deciding what to interpret. Learn about the land's history and today's Tribal communities who are connected to it. An interpreter's enthusiasm and understanding are infectious.

- Learn about the audience: Will schoolchildren attend? Retired people? International visitors? Are they likely to be familiar with spending time outside? What might interest them? What do they need or expect? Will you be presenting in their first language? Adapting a program to an audience leads to success.

- Explore meanings of the topic, especially universal ones: a desert can represent an ancient seasonal migration route, a deadly wasteland, a complex web of life, or a source of mineral wealth. These larger themes help people connect to the value of nature and unique places.

- Develop a theme—a main idea—that takes into account the topic, the audience, and the meanings: the theme of the program provides justification for why the audience should care about the resources and surroundings.

- Choose techniques that make the theme come alive: engaging all five senses and using props, quotes, stories, activities, direct observations, and other techniques help people connect with the topic and theme.

- Organize the program around the theme: Choose secondary ideas to support the main idea. Identify potential transitions to move from one idea to the next.

- Practice the presentation, and over time it will improve: After each run-through, explore what worked and what didn't, and continue improving the experience to keep it fresh. Find answers for questions you were asked that you couldn't answer—a great naturalist loves to say "I don't know," but not twice to the same question!

It's also important to be sensitive to the trauma people may have experienced when you cover topics such as fire, climate change, drought, or even contact with animals. Begin by creating a safe and welcoming environment, acknowledging the potential presence of trauma, and respecting individuals' experiences and feelings about these topics. Some people may need alternative ways to participate so they can comfortably engage with the material being shared. Recognize that these experiences or even just being outside away from familiar places and the built environment can evoke various emotions and reactions.

Interpretation can take many forms, both formal and informal, and can include nature walks, visitors' center desk duty, campfire talks, school trips, and historical reenactments. Other forms of interpretation include roving on trails and talking with visitors at popular overlooks. This approach contrasts with more-formal talks and walks. The interpretive talk and the naturalist walk are two of the more common experiences where naturalists serve as interpreters.

An Interpretive Talk

An interpretive talk—a prepared presentation usually lasting 10 to 20 minutes—can be given in or near an interpretive center, at a campfire, or in a similar venue. Interpretive talks require thorough planning,

A naturalist informs class participants about wetland plants. Photo by Kerry Heise.

practice, and feedback from colleagues and audiences. Enjoy the preparation! We learn by teaching, and preparing a talk is like preparing for an engaging class. Plan what to say, and how to say it. What theme will capture the audience's interest? What interpretive techniques will make it come alive? Also, going to other interpreters' programs can help you identify what works and what doesn't and can inspire new ideas.

Let's walk through an example of one way to prepare an interpretive talk, based on the process mentioned above. Naturalists at a state park might choose to do a talk on giant sequoias, the world's most massive trees. They first spend time walking among and observing the giants, and they listen to park visitors to hear what intrigues them and what questions they have. They do library and online research and talk with colleagues to get a broader picture of sequoias' history and ecology. They learn about people who have inhabited the place and how they relate to these trees, placing the people and the land in context.

The naturalists then review the park's or organization's mission and think about what the trees have meant to people through time: a source of wonder and mystery, a symbol of survival, the target of get-rich-quick lumber schemes, the harmful removal of Indigenous Peoples and the resilience of Tribal communities today, and more. Sequoias' size and survival ability fascinates visitors, so the naturalists develop a theme based on this: sequoias survive long odds to become some of the biggest and oldest beings on Earth.

How can the naturalists help an audience connect with this idea? They might bring tiny sequoia seeds—the size of a Quaker oat flake—to illustrate how hard it is for a sequoia to survive its first year. They could talk about how many hundreds of gallons of water sequoias "drink" every day during the warm months to survive, and how drought makes this difficult. They might show pictures or point out fire scars to help people imagine how sequoias' thick bark protects them from fire.

The naturalists then organize a talk around the selected theme and related ideas. They plan ways to adapt it for different audiences. For adults, they might address how climate change could hamper sequoias' survival. For children, they might add a game that shows how sequoias depend on wind, squirrels, and fire to spread their seeds. Then after practicing the program with a few friends or colleagues, they are ready to try it on the public and continue to adapt, depending on the responses they get. A talk lasting 10 or 20 minutes should contain one big idea, or theme. This big idea should be supported by a number of smaller ideas (subthemes) presented throughout the talk that relate back to the original big idea. When naturalists circle back and reemphasize the big idea behind various subthemes, the audience hears it multiple times and has a series of opportunities to grasp it. That ensures that they have an important message or better understanding of an important theme to take home with them.

As they do in other situations, some people at interpretive programs will pay more attention to nonverbal cues than to words. So let your passion for what you are presenting come through—let your light shine. It is the beacon that will tell people "this stuff is so cool—check it out!"

A Naturalist Walk

Naturalist walks are like interpretive talks in terms of preparation. They often have designated topics, such as birds, banana slugs, or global change, and they usually have themes. Interpreters must know their topics, their audiences, and their routes really well. Preparing for an outing requires a few other logistics, but don't worry! You have the best props and stage hands in the world—nature!

In thinking through the walk, plan stops, techniques, and transitions. For example, one stop might be a good place to explore feeding behavior by using binoculars to watch birds eating ripe berries, and another might provide an opportunity to examine tracks of bears and coyotes. These stops might shift with the day or the group. Transitions help the group move between stops. For example, questions lead people to look closely or think about a new idea as they walk from one place to the next. It's also important to encourage the group to be silent through parts of the walk so everyone can focus on the natural sounds and subtle interactions happening around them.

In guiding a hike, it's useful to have a few "elevator speeches"—minute-long talks about something commonly seen along a trail. These provide an opportunity to generalize a specific observation, make meaning of an experience, or connect facts to big ideas. For instance, a question about a paintbrush flower (*Castilleja*) might allow an interpreter to tie that flower to the broader idea of pollination, weaving in the fact that hummingbirds are attracted to the color red and that red paintbrushes are pollinated by hummingbirds.

On the day of the walk, arrive early and get to know the audience. This can help in tailoring the walk to the group. At the start time, welcome people, introduce yourself, and perhaps mention reasons you love the particular place you are walking. Give a brief description of the program so that the audience knows what's coming.

It's important to keep safety and comfort in mind. Thirst, heat, and cold are conditions you must keep tabs on because they will sap attention as effectively as a boring lecture. If people are physically uncomfortable, they won't be able to take in the experience. Part of a standard introduction might be to remind everyone to carry water and layers of clothing they can add or subtract, then allow them time to arrange their gear. Let the group know where the restrooms are and how long it will be until the next bathroom stop.

In addition, make the audience aware of safety concerns that might crop up. Safety first! For example, poison oak is so common in California that it should be written into the naturalists' job description that on each hike they show the plant to people and explain how to recognize it. If asked, people will often say they know poison oak, but it's wise to assume no one has ever seen the plant before, especially in the late fall and winter when it doesn't have leaves.

Rattlesnakes and ticks are other common hazards that naturalists should address in California. Learning how to recognize and deal with

A young naturalist participates in the City Nature Challenge in Sacramento. Photo by Ryan Meyer.

rattlesnakes protects both people and snakes. Good rules of thumb include watching where you put your hands, feet, and rear and giving snakes a wide berth rather than trying to kill them. Be sure to let the group know guidelines for an area, such as staying on the trail, acceptable noise level, and whether it's appropriate to pick flowers or take anything from the site.

During the walk, regularly take the pulse of the group. Doing this can be challenging at first, but it's worth the effort. Are they done with this flower? Have they spent enough time walking? Are they tired and in need of water or shade? Watch people's body language. Are they making eye contact? Are they ignoring the program or wandering away physically or mentally? Are they falling behind as the group ascends a hill? Are they fidgeting? Asking people how they feel is legitimate, and they will usually respond if asked.

On a walk, let the world be the guide. If the group comes across a frog or a beetle in the path, stop and observe it. Let people experience it and then talk about it. Sharing an experience and talking about it will be more meaningful to the group than knowledge alone. When the group is observing an object, excellent naturalists keep the focus on the object, rather than on themselves.

On guided walks people often ask, What's the name of that thing (bird, bug, flower)? Naming gives the impression that a thing is known, when it may really allow participants to not observe closely! Instead of answering and moving on, try turning the naming question into a question that requires observation and encourages people to look more closely. When asked the name of a flower, a naturalist might respond with, What do you notice about that flower? or Are all the petals the same? If the question is about a bird, consider asking, How many colors can you see on the bird? or suggesting, Let's just listen for a few minutes so we can all hear this bird's song. Extending the observation allows more people to be drawn into the experience and to make it their own.

Well-trained spokespeople in the media often avoid answering questions or talk around them. In contrast, naturalists usually serve best when they answer questions directly. Sometimes this means cultivating the ability to say "I don't know." The natural world is bigger, more diverse, and more complex than any one person can comprehend. "I don't know" doesn't mean "end of story, let's move on." It can mean "we're all beginners here, so what questions can we ask?" It might mean "this is a new experience; isn't that wonderful?" Or it could mean "I'm not the only authority on this walk; perhaps someone else can add something here." Offer ways that people can discover an answer. Invite them to enjoy the idea that nature is a mystery. Or use the question to draw connections between what they're observing and the big ideas in the program.

Saying "I don't know" can also be followed with a suggestion on how to follow up and find out. Taking a questioner's name and email address, finding the answer to the question, and sending it reconnects the questioner to the experience and often leaves a lasting impression. Typically, interpreters who lead walks repeatedly in one place are asked a finite set of questions over and over. These questions inform them of the audience's interests. Jotting down questions and then learning the answers deepens knowledge and experience for both you and your audience.

Using Social Media for Natural Resource Interpretation

A social media post is a type of short-form presentation. While posts do not fit within the traditional categories of an interpretive talk or naturalist walks, they can be an effective way to reach a broad audience about a naturalist topic. Here are a few tips for effective use of social media:

- You are reaching out to a saturated and distracted audience. Keep it short. Keep it targeted. One topic at a time!
- Funny, engaging, and cute posts are always good.
- Use visuals. Wildlife camera footage and other videos are popular. Time series photos, videos, or maps can show transformations effectively (think of the photos that show changes to snowpack year to year).
- Consistency is essential. Because of the algorithms employed by most platforms, users who don't post regularly will get lost among the torrent of daily content.
- Partner with other organizations to boost your promotional efforts, and use hashtags to connect to existing efforts.

COLLABORATION AND CIVIC ENGAGEMENT

Beyond interpretation, naturalists use many other approaches for communicating ecological information. They talk with decision-makers about policies, organize action groups, and get help creating and implementing restoration projects. While technical solutions are essential for solving ecological problems, perhaps the most important skills for protecting and restoring landscapes are interpersonal ones.

Collaboration in the Community

Collaborating with other community members can be a powerful tool for ecological change. Speaking to neighbors or city councils, organizing field trips, getting information from resource agency personnel, or making decisions as part of a watershed group are all facets of becoming an active steward and building effective community relationships. Working with an existing group or forming a new group to work on an environmental issue of concern is often the most effective way to achieve a workable solution.

Speaking at Public Forums

The following tips can guide successful communication at a formal or informal meeting:

- Arrive before the meeting starts and sign up to speak, if necessary.
- Prepare your comments ahead of time, making sure they fit into the time limit.
- Be specific about your reasons for supporting or opposing proposals or actions.
- Try to be calm while still being passionate.
- Stay positive! Talk about the desired outcome, rather than assigning blame.
- Be willing to listen to opposing points of view. Identify common interests and areas of agreement.
- Start with an assumption of goodwill on the part of decision-makers and agency personnel. Most people are doing their best to do their job and assist the public.
- Stay for all of the relevant part of the meeting rather than leaving after you make a statement. Listen to others and be available for discussion.

Informing the Community about a Project

To encourage others in the community to learn about or participate in a project or action, or to start one of their own, consider these ideas:

- Invite the public to tour your project.
- Inform local agencies about your project, and see if they can help or direct you to additional resources. Many agencies are underfunded, and so staff time may be limited. Be succinct and patient in your communication with them.
- Write an article about your project for a local newsletter or newspaper, or start a blog about it.
- Give out contact information or brochures offering assistance with similar projects.
- Provide information on a website.
- Use social media to inform people about new developments and encourage participation in events.

Running Effective Meetings

Watershed groups, planning committees, and resource councils are effective forums for working with a variety of interested parties to meet common goals. When done well, meetings can be efficient places to share ideas and get business done. The following tips are for running productive meetings:

- Draw up and communicate a realistic agenda, and then stick to it.
- Determine ahead of time how decisions will be made.
- Learn whether a formal structure (such as Robert's Rules of Order) will be required or desired for meeting discussions.
- Agree on rules for courteous discourse.
- Rotate the leader/facilitator role among regular participants in an ongoing project.
- Assign a timekeeper and a notetaker.
- If visuals are used for a meeting, such as a PowerPoint presentation, be sure that text, maps, graphs, and other graphics are large enough to easily be seen by someone at the back of the room.
- End meetings on time. This is very important, to encourage ongoing participation.

Collaborative Conservation

Besides talking with the public at large, naturalists also may participate in focused communication on behalf of the environment. Collaborative conservation brings together diverse groups and people who work together, trying to develop solutions that address the needs and perspectives of everyone involved. It can be an important first step in planning for a natural resource project or policy, especially one involving a wide range of opinions and strong feelings. Interested parties in these projects will include local government, Tribes, land management agencies, neighbors, commercial users (timber, livestock, agriculture), recreational users (hiking, skiing, fishing, hunting), local or state conservation groups, and possibly real estate developers and nearby business owners. Boards, councils, or other decision-making bodies should be represented as well. Unless all the interested parties are involved from the beginning, some members of the community may feel sidelined.

Collaborative conservation often requires public participation and frequently falls into one of two kinds of efforts: community based and policy based. A community-based project involves people who share the same place or are part of the same community, or a group of people who share a desire to address specific policies or interests. A local watershed group might involve members who live in the watershed but also members who recreate but don't live there. Similarly, Tribes and Tribal communities may have an interest in a project within their historical homelands. A policy-based project, on the other hand, might involve government agencies or officials working to develop a regional plan or local ordinance.

There is some evidence that community-based efforts work best when they're led by local participants and when the process is open and includes and accommodates everyone's perspectives. This includes government representatives. Groups focused on smaller areas are more likely to succeed. Those involved can relate to the landscape in question, and regular participation from people spread across a large geographic area is not required.

The Quincy Library Group in Northern California is an illustrative example of a good faith effort that faced challenges we can learn from. An approximately 30-person steering committee developed a plan for 2.5 million acres of public forestland. Though laudable for the group's attempt to try to work cooperatively and include multiple stakeholders, in the end the process faced many challenges, in part because of the many interests in this large and relatively populated area. Confusion, failure, and even damage can result when individuals are held accountable, or hold themselves accountable, for large and diverse interest groups. Such larger-scale conservation projects are better addressed through a network of local efforts.

No matter the format, making the objectives clear and agreeing on goals are important first steps for any group process. The goals should be specific, and they should be identified individually even if multiple goals overlap. Doing this requires being clear from the start about what the group has the ability and authority to accomplish. For example, a Groundwater Sustainability Agency may convene interested parties to decide how to implement groundwater recharge projects, but they do not have the ability to change the state law that compelled them to develop a Groundwater Sustainability Plan in the first place.

Also, getting everyone on the same page about what information is available is critical. One way to do this is to collect and evaluate exist-

Tending the living roof at the California Academy of Sciences, San Francisco. Photo by Susan Mcstravick.

ing information and share it with the group. This will increase everyone's understanding of the problem and possible solutions and provide opportunities for people to test their hypotheses. And it's a good way to empower the group early on.

Presenting preliminary ideas to different interest or expert groups at various times during the planning process for feedback can also be an effective way to involve interested parties. In the case of emotional issues that may polarize a community, this approach can keep opposing interest groups from derailing the process. Keeping all parties informed, including the public, is essential for buy-in and to prevent, or at least minimize, discord. Collaborative conservation plays an important role in later stages of discussion too: the coalitions that can arise are often in a good position to respond to future disagreements or policy disputes, perhaps preventing future crises from arising. In summary, successful collaborative conservation efforts can increase community resilience to challenging environmental issues.

Environmental Stewardship

Stewardship is an action-oriented ethic focused on caring for the natural world. Stewardship has its basis in natural history, among other disciplines, and is the practice of promoting sustainability, biodiversity, and resilience. As an approach to inhabiting the Earth, it is similar to

Native American principles of reciprocity, respect, and relationship with the entire community of life. Stewardship is realized through actions shaped by place, scientific knowledge, traditional knowledge, and public participation, often in the form of ecosystem restoration and resilience, biodiversity conservation, and enhancement of ecocultural resources. Connecting with others on actions that tend the land is also a path to joy and hope.

Here are a few ways you and your community can join together to practice environmental stewardship:

- Contact Cal PBA, California's hub for Prescribed Burn Associations (PBAs), which can connect you to a local PBA that relies on volunteers to lay good fire on the land to improve the health of ecosystems and protect communities. Prescribed Fire Training Exchanges (TREX) offer members of the public experiential training in prescribed fires and ways to connect with others engaged in beneficial burning.

- Volunteer for the California Invasive Plant Council, which offers information and opportunities to inventory and monitor invasive plants as well as manage them.

- Join a California Native Plant Society local chapter and go on a naturalist-led hike or assist with native plant sales or rewilding.

- Join a river friends group, such as Friends of the Los Angeles River or Friends of Gualala River, to help restore streams and wetlands.

- Restore habitat in a protected area or wildlife corridor with a park friends group like Amigos de Bolsa Chica, California Trout, Orange County Coastkeeper, or the Golden Gate National Parks Conservancy.

- Reach out to California ReLeaf to find a community-based group engaged in tree planting and care, such as TreePeople in Southern California.

PARTICIPATORY SCIENCE

Participatory science is an exciting and growing method of collecting scientific data in which the general public participates in scientific studies ranging in scale from local stream monitoring to bird counts spanning the globe. It was once commonly referred to as citizen science, but

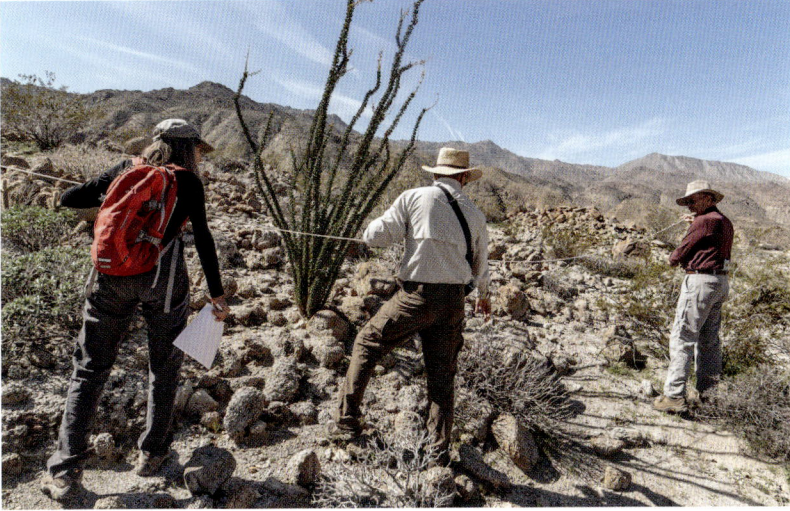

Volunteer naturalists gathering data near the Randall Henderson Trail within the boundaries of the Santa Rosa and San Jacinto Mountains National Monument. Data collection is related to a nearby ocotillo (*Fouquieria splendens*), pictured, under direction of researchers at the University of California, Riverside, Center for Conservation Biology. Photo by Colin Barrows.

we refer to it as participatory science to emphasize inclusivity, and sometimes as community science to emphasize a collaborative approach. For example, the Community Collaborative Rain, Hail, and Snow Network (CoCoRaHS) of the National Weather Service wants you to collect data in your own yard or on your rooftop! These studies take advantage of having tens to thousands of observers in the field contributing important data. Today, bird-watchers have become essential for studying changes in bird abundance and distribution by collecting data during the National Audubon Society's annual Christmas Bird Count. Many anglers are experts in identifying stream insects and contribute to projects monitoring invertebrate populations and other indicators of water quality. The California Bumble Bee Atlas has a huge group of volunteers collecting data on bumblebee and other native bees. These projects allow scientists to gain a greater understanding of how, for example, water quality may change across a watershed, or how a warming climate may affect bee populations.

Members of the public contribute to participatory science for many different reasons—it's fun to engage in scientific research, it deepens personal knowledge and connection to the Earth, it uses observations people are already making in nature, it helps society better understand nature, and it provides a growing body of scientific data.

Besides aiding scientific studies, community contributors learn key scientific concepts and natural history firsthand. They experience the excitement of doing scientific research, which fosters their continued interest. Participants often become experts at identifying organisms or measuring environmental factors, such as air quality, and they can document ongoing changes in local plant and animal populations. Participants also build social bonds with each other, which may lead to becoming more active in making positive changes in their communities because of the experience they gain in carrying out research.

Volunteer monitoring programs engage people with a wide range of experience and knowledge, from people with little scientific training to naturalists with greater local knowledge than some academics. Projects may be "contributory," in which participants follow protocols developed by scientists; "collaborative," in which participants inform and improve research design and disseminate findings; or "co-created," in which the public helps generate research questions, analyze data, and distribute results.

The range of opportunities to participate in research is growing, and new online communication tools, including apps and cell phones, are being used by the public to help collect data. Many projects need input and would welcome the participation of California Naturalists. See the Explore! box for some ways to start.

Being a naturalist offers a lifetime of experiences connecting to the community of life. By embracing this role, you have the power to make a tangible difference in protecting the Earth. Earth stewardship can seem overwhelming, especially under the pressure of climate change. Connecting with others provides an opportunity for transformation and a path to hope. We invite you to use the California Naturalist course as the beginning or the next step of your journey of collaboration, communication, lifelong learning, and personal engagement in the natural world. Let curiosity and kindness be your guide. When you act, you will find others to join you. We can do this, and the only way forward is together.

Explore!

VISIT A NATURE CENTER

Most nature centers have well-established interpretive programs that can provide examples of success. Nothing beats seeing experts at work for inspiration.

VISIT A HISTORIC INTERPRETIVE SITE

These sites often have well-developed living history programs that can offer innovative ways to present programs.

JOIN A LOCAL COMMITTEE

Take a front seat at local planning decisions! Most cities and counties have dozens of natural resource advisory groups on topics such as parks and open space, water resources, or wildlife. Committees and commissions may require appointment by the city council, board of supervisors, or the like, and in many places there are more vacancies than interested members of the public.

PARTICIPATORY SCIENCE

Participate in scientific research! You can select from a wide range of projects on the UC Environmental Stewards website, or take an online Participatory Science training course. For more information, see the UC Davis Center for Community and Citizen Science web page.

Glossary

ABIOTIC Physical rather than biological; not derived from living organisms; for instance, rocks, water, and air.

ACCRETIONARY WEDGE A mass of sediments and oceanic lithosphere scraped off the top of a subducting oceanic crustal plate and onto the less-dense, nonsubducting plate at a convergent plate boundary.

ACID PRECIPITATION Rain or any other form of precipitation that is unusually acidic, mostly caused by human emissions of sulfur and nitrogen compounds that react in the atmosphere to produce acids.

ADAPTATION The process that results in an organism becoming better suited to its environment and improving its survivorship over successive generations. Can also be a characteristic that is important for survival (as in *an adaptation*).

ADVENTITIOUS ROOTS Roots that come from stems or other plant parts, but not from the main root. They are handy for climbing or holding on; for example, ivy climbs using adventitious roots.

ALLUVIAL Describing sediment that is or has been carried by water. See also *colluvial*.

ALLUVIAL FAN A geological feature created when a steep, confined stream leaves a canyon, widens out, and drops a portion of its sediment, creating a broad deposit.

ALLUVIAL VALLEY A valley along a stream or former stream where most of the upper layers of the soil have been deposited by alluvial processes.

ALTERNATE LEAVES An arrangement of leaves (or buds) on a stem in which only one leaf arises from a node, creating the appearance of leaves stepped (alternating) on the stem. See also *opposite leaves* and *whorled leaves*.

AMNIOTIC EGG A fluid-filled egg in which the developing embryo is protected by a series of membranes and often a hard or leathery shell that resists desiccation.

AMPHIBIANS Ectothermic vertebrate animals whose eggs do not have an amniotic membrane. Usually four-limbed, and usually with a juvenile water-breathing form (tadpole) that metamorphoses to an adult air-breathing form. Includes frogs, toads, salamanders, and caecilians.

ANADROMOUS Describing fish that spend part of their lives in the ocean and migrate to fresh water for breeding.

ANNUAL A plant that lives for only one growing season. See also *perennial*.

ANOXIC Without oxygen, or having dangerously low levels of oxygen.

ANTENNA One of the jointed, movable, sensory appendages occurring in pairs on the heads of insects and most other arthropods except spiders and scorpions.

ANTHROPOGENIC Relating to or resulting from the influence of human beings.

AQUATIC In water. Aquatic plants and animals live in water.

AQUIFER An underground layer of porous and permeable rock or soil that holds groundwater.

ARTHROPOD An invertebrate animal with bilateral (left-right) symmetry, a hard exoskeleton, jointed legs, a segmented body, and many pairs of limbs. Arthropods include insects, spiders, centipedes, shrimp, crayfish, and many others.

ASPECT The direction that a hill or mountain slope faces (such as north, south, east, or west).

ATMOSPHERE The gaseous envelope surrounding the Earth; the air.

ATOM The smallest component of an element having the chemical properties of the element.

AUTOCHTHONOUS Describing anything that originates in or is native to the area in which it is found.

AUTOTROPH An organism capable of generating its own food; generally, plants that produce food through photosynthesis.

BAJADAS Depositional areas at the mouths of canyons where steep slopes merge with flats and the sediment carried by the creek is spread in a fan-shaped apron.

BEDLOAD Macroscopic particles of rock and mineral carried by a stream or other body of water. See also *suspended load*.

BEDROCK Solid rock underneath loose material such as soil, sand, clay, or gravel.

BEHAVIOR The actions or reactions of organisms in response to external or internal stimuli.

BINOMIAL NOMENCLATURE The system of two-part Latin names used to give each species a unique name consisting of a combination of a genus and a species name.

BIODIVERSITY The number and variety of organisms within a given ecosystem or geographic area.

BIOFUEL Fuel produced from renewable resources, especially plant biomass, vegetable oils, and treated municipal and industrial wastes.

BIOGEOCHEMICAL CYCLES The movement of matter through Earth's systems in which chemicals or molecules move through both biotic and abiotic stages, as they do in the water cycle, the carbon cycle, and the nitrogen cycle.

BIOGEOGRAPHY The study of the distribution of biodiversity over space and time. It aims to reveal where organisms live, and at what abundance.

BIOMASS The mass of living biological organisms in a given area or ecosystem at a given time.

BIOREGION A region defined by characteristics of its natural environment, such as flora, fauna, climate, habitat type, and topography.

BIOSPHERE The area on Earth in which life exists.

BIOTA The total of all organisms within a given ecosystem at a given time.

BIOTIC Biological; derived from living organisms.

BODY PLAN The pattern or blueprint for the way the body of an organism is laid out.

BRACT A leaf in a flower cluster or a leaf base of a flower, usually differing somewhat from an ordinary leaf in size, form, or texture. It is often much reduced but occasionally large and showy, sometimes petallike, highly colored, and very conspicuous.

BROADLEAF HARDWOOD A tree with wide, flat leaves, as contrasted with needles, that is usually deciduous, with examples including oak, cottonwood, and alder. The wood is denser and hence harder than the wood of most conifers.

BRUSH Shrubland; a habitat composed of many-stemmed woody plants less than 15 feet tall. See also *chaparral*.

BRYOPHYTES Plants that have no vascular tissues and reproduce by spores, not flowers. They are usually very small. Bryophytes include mosses, liverworts, and hornworts.

BUD A growing point enclosed by closely overlaid rudimentary leaves. The tissue that will become a stem, leaf, or flower.

BULB A swollen underground vertical shoot that has modified leaves (or thickened leaf bases) that are used as food storage organs by a dormant plant.

CAMOUFLAGE Protective coloring or another feature that conceals an animal and enables it to blend into its surroundings.

CARBON EMISSIONS The release of carbon into the atmosphere over a specified area and period of time.

CARBON SEQUESTRATION The process of removing carbon from the atmosphere and depositing it in a reservoir. Carbon sequestration naturally occurs during photosynthesis.

CARNIVORE An animal that feeds primarily on the meat of other animals rather than plant sources.

CARTILAGINOUS Having a skeleton consisting mainly of cartilage, a type of dense connective tissue composed of specialized cells called chondrocytes. The skeletons of sharks, rays, and skates are made of cartilage.

CASCADE RANGE The primarily volcanic mountain range running north-south from British Columbia, Canada, south through Washington and Oregon and into California.

CELLULOSE A stringy, fibrous structural substance in the cell walls of plants.

CENTIPEDES Predacious arthropods with elongate, flattened bodies composed of 15 to 173 segments, each with one pair of legs, the first pair being modified into poison fangs.

CENTRAL VALLEY The long, wide valley in the center of California between the Coast Ranges to the west and the Sierra Nevada to the east. Sometimes called the Great Valley.

CEPHALOTHORAX The anterior section of arachnids and many crustaceans, consisting of the fused head and thorax.

CHANNELIZATION Activity (usually human) that moves, straightens, shortens, cuts off, or diverts a stream channel, whether natural or previously altered.

CHAPARRAL A plant community common to California and other Mediterranean climate regions characterized by dense evergreen, drought-resistant shrubs and small trees such as manzanita, scrub oak, chamise, and ceanothus. See also *brush*.

CHELICERAE The first pair of usually pincerlike appendages of spiders and other arachnids.

CHEMICAL ENERGY Energy liberated by a chemical reaction or absorbed in the formation of a chemical compound.

CHEMOSYNTHESIS The making of organic compounds using energy derived from chemical reactions, typically in the absence of sunlight.

CHITIN A tough, protective, semitransparent molecule. Chitin is the main structural component of arthropod exoskeletons and the cell walls of some fungi.

CHRYSALIS The pupal case of many insects, especially of moths and butterflies, in which metamorphosis takes place; the cocoon from which the adult eventually emerges.

CIRQUE A shallow horizontal depression formed by glacial action, often holding a small lake.

CISMONTANE CALIFORNIA Land to the west of the Sierra-Cascade mountains.

CLEAR-CUTTING A tree harvest technique involving felling and removal of all trees from a given tract of forest.

CLIMATE The combination of all weather factors, such as temperature, humidity, and precipitation, over a long period of time.

CLONE A plant or other organism that has reproduced asexually to produce genetically identical offspring. Hundreds of aspen stems may be clones of one original seed, which may have germinated thousands of years ago!

CO2 (CARBON DIOXIDE) A chemical compound composed of two oxygen atoms bonded to a single carbon atom. It is a gas at room temperature and pressure and exists in Earth's atmosphere as a gas. Many vertebrates breathe out carbon dioxide.

COAL-FIRED POWER Power produced by burning coal, which is used to generate electricity.

COAST RANGES The group of mountain ranges more or less directly inland from the Pacific coast, running north-south from Northern California to Ventura County in the south and making up the western border of the Central Valley.

COCOON The silk envelope spun by the larvae of many insects, serving as a covering while they are in the pupal stage of metamorphosis.

COLLABORATIVE CONSERVATION A deliberate and inclusive process of two or more people, groups, or entities coming together to work out issues related to sustaining and improving natural resources and human communities.

COLLUVIAL Describing sediment that was moved by gravitational processes rather than water, like landslides. See also *alluvial*.

COMMUNITY Populations of different plant and animal species occupying the same geographical area.

COMMUNITY CHOICE AGGREGATION A program that allows local governments located within the service areas of investor-owned utilities such as PG&E or Southern California Edison to purchase and/or generate electricity, often green electricity, for their residents and businesses.

COMPLETE METAMORPHOSIS Insect development in which the initial stage looks remarkably different from the adult. Egg, larval, pupal, and adult stages occur, each differing greatly. Beetles, bees, flies, and butterflies use complete metamorphosis.

COMPOST A mixture of various decaying organic substances, such as dead leaves or manure, used for fertilizing soil.

COMPOUND LEAF A leaf that has a fully subdivided blade, each leaflet of the blade separated along a main or secondary vein. See also *simple leaf*.

CONE OF DEPRESSION A decrease in water pressure or lowering of the water table that occurs when groundwater is pumped from a well.

CONFINED STREAM A stream with bedrock canyon walls or other obstructions marking its banks and preventing lateral changes to its course.

CONIFER A plant bearing seeds in cones rather than flowers. Conifers include pine, fir, redwood, and spruce.

CONNECTIVITY A measure of the extent to which plants and animals can move among habitat patches. Landscape features such as corridors, greenbelts, and ecological networks provide potential means for retaining habitat connectivity.

CONSERVATION The protection, preservation, management, or restoration of species or natural environments. Conservation is generally held to include the management of human use of natural resources for current public benefit and sustainable social and economic utilization. Conservation can be in situ (where the organism lives) or ex situ (in zoos or botanical gardens).

CONSERVATION EASEMENT OR COVENANT A deeded transfer of partial interest in real property to a private or public institution to conserve land or its resources for future generations. Conservation easements can result in tax benefits for landowners and are binding on all future owners of the property.

CONSUMER An organism that derives its energy from other organisms or organic matter.

CONVERGENT PLATE BOUNDARY A tectonic boundary where two plates are moving toward each other.

CORE The dense innermost portion of the Earth composed mostly of iron, nickel, and other metals.

CORE HABITAT Habitat that is far enough away from other habitat types to avoid their influences. See also *edge habitat*.

CREPUSCULAR Describing animals that are primarily active during twilight, that is, at dawn and dusk.

CROPLAND Land that is managed by humans for intensive agriculture.

CRUST The outermost portion of the Earth comprising the continental surfaces as well as ocean floors and varying in depth from 3 to 40 miles.

CRUSTACEAN Any of various predominantly aquatic arthropods of the sub-phylum Crustacea, including lobsters, crabs, shrimps, and barnacles, charac-teristically having a segmented body, a chitinous exoskeleton, and paired, jointed limbs.

DAM An artificial blockage along a stream created for the purpose of impound-ing water.

DEAD ZONE A low-oxygen area in an ocean or lake due to an excess of nutri-ents, typically from fertilizer runoff, and resulting in a large reduction in aquatic life.

DECIDUOUS Describing a plant that sheds foliage at the end of the favorable season.

DECOMPOSER An organism, often a bacterium or fungus, that feeds on and breaks down dead plant or animal matter, thus making organic nutrients available to the ecosystem.

DECOMPOSITION The breakdown of a formerly living entity into simpler materials, minerals, and nutrients that provide sustenance to other organisms.

DELTA The end of a stream where it flows into an ocean or other large standing body of water allowing the stream to deposit its suspended sediment over a widening area.

DENTICLE A toothlike or platelike scale on the outside of a cartilaginous fish, such as a shark or ray.

DEPOSITION ZONE A portion of a stream where large amounts of suspended sediments are deposited on the banks and/or bed.

DESERT A climate region that receives less than 10 inches of rainfall annually.

DESICCATION The drying out of a living organism.

DETRITIVORE An organism that feeds on and breaks down dead plant or ani-mal matter, returning essential nutrients to the ecosystem.

DIURNAL Active during the day and sleeping at night.

DIVERGENT PLATE BOUNDARY A tectonic boundary where plates split apart or separate from one another and allow molten mantle material to rise up.

DIVERSION Removal of water from a stream, usually for human use.

DIVERSION CHANNEL A man-made canal that carries water from a stream and delivers it elsewhere, often for agricultural use.

DOMESTICATED Describing animals that are tamed, especially by generations of breeding, to live in close association with human beings as pets or work animals and that usually have been made dependent so that they lose their ability to live in the wild.

DORMANT Being in a state of rest or inactivity; quiescent.

DRAINAGE BASIN The topographic region drained by a river and its tributar-ies. Also called a catchment basin or a watershed.

DYNAMIC EQUILIBRIUM The stability of a system in a steady state. Change is constant, but balance is restored.

ECOSYSTEM The combination of all living organisms and natural features of a given geographic area, with emphasis given to their interdependence.

ECOSYSTEM MANAGEMENT A management approach that includes all the organisms and abiotic factors that affect the area. A contrast would be single-species management.

ECTOTHERMIC Referring to organisms whose body temperature varies depending on external conditions. Compare *endothermic*.

EDGE HABITAT Habitat on or near the boundary between two different habitat types. See also *core habitat*.

ELAIOSOME A fleshy, protein-rich "food patch" on some seeds or fruits that rewards dispersers. Ants are "paid" with elaiosomes to move seeds.

ELECTRICAL ENERGY Energy made available by the flow of electric charge through a conductor.

ELECTROMAGNETIC ENERGY A form of energy that is reflected or emitted from objects in the form of electrical and magnetic waves that can travel through space.

ELECTROMAGNETIC WAVE A wave of energy having a frequency within the electromagnetic spectrum and propagated as a periodic disturbance of the electromagnetic field when an electric charge oscillates or accelerates.

EMBODIED CARBON The unseen greenhouse gas emissions associated with a product's entire life cycle. It includes emissions released in extraction, production, transport, and manufacturing of a product. Think beyond how much greenhouse gas you save by using solar panels; think of the emissions released in the creation of the panels.

EMBODIED ENERGY The energy that was used to make a product and to get it from its source to the consumer.

ENDEMIC Describing a species (or other well-defined group) that is native only to a single geographical area. See also *paleoendemic*.

ENDOTHERMIC Describing organisms that maintain relatively constant body temperatures regardless of the outside temperature. Compare *ectothermic*.

ENVIRONMENTAL JUSTICE A movement that affirms the right of all people to environmental protections and benefits and works to remedy historic injustices to poor or underserved communities, such as disproportionate exposure to pollution, poor air and water quality, and underinvestment in infrastructure and green spaces.

EPICORMIC SPROUTING Growth that emerges from dormant buds along the trunk or branches of a tree.

EPIPHYTE A plant that stands or hangs on or is supported by another plant. Epiphytes are not parasites; they derive nutrients from rain.

EPOCH A unit of time in geology, the time during which a rock unit was deposited. Epochs typically cover a few million to several tens of millions of years. Examples are the Holocene, Pleistocene, and Miocene.

EROSION The removal of soil and rocks from the surface of one area and their deposition elsewhere by processes such as gravity, water flow, wind, or glacial action.

ESTIVATION Also known as summer sleep, a state of animal dormancy somewhat similar to hibernation. Animals are known to estivate to avoid high temperatures and the risk of desiccation.

ESTUARY The place where a river meets the sea. An aquatic coastal habitat with both freshwater and marine influences.

ETHANOL A volatile, flammable, colorless liquid best known as the type of alcohol found in alcoholic beverages and in modern thermometers.

EUKARYOTE Any organism having cells with internal membrane-bound organelles, especially the nucleus. Eukaryotes include most of the multicellular life-forms you are familiar with. Contrast prokaryotes, bacteria, blue-green algae.

EUTROPHIC Describing a water body having an abundance of mineral nutrients, particularly nitrogen and phosphorous, often due to fertilizer or animal waste runoff.

EUTROPHICATION A process in which water bodies receive excess nutrients that stimulate excessive plant growth, especially of algae.

EVERGREEN A plant that retains its leaves for one or more years; having leaves all year-round. This contrasts with deciduous plants, which lose their foliage for part of the year.

EVOLUTION The idea that all species are related and change gradually over time as a result of adaptations to their environments; change in a population during successive generations, as a result of natural selection acting on genetic variation among individuals, and resulting in the development of new species.

EXOSKELETON A hard, protective outer body covering of an animal, such as an insect, crustacean, or mollusk. The exoskeletons of insects and crustaceans are largely made of chitin.

EXOTIC SPECIES Species of organisms introduced into habitats where they are not native.

EXTINCTION No longer being in existence. A species is said to be extinct if it has been killed off or died out.

EXTIRPATED Locally removed. A species (or other taxon) is said to be extirpated if it no longer exists in the chosen area of study but still exists elsewhere.

EXTRUSIVE Describing igneous rock that cools and solidifies aboveground. Contrast *intrusive*.

FERTILIZER Any of a large number of natural or synthetic materials, including manure and nitrogen, phosphorus, and potassium compounds, spread on or worked into soil to increase its capacity to support plant growth.

FISHERY Harvesting of fish, shellfish, and sea mammals as a commercial enterprise; also the location or season of commercial fishing.

FLOODPLAIN The portion of land bordering a stream that receives periodic flooding.

FLOWER The reproductive organ of a seed plant, typically made up of sepals, petals, stamens, and the pistil, with the most visually striking parts often being the petals.

FOOD WEB A series of organisms related by predator-prey and consumer-resource interactions; the entirety of interrelated food chains in an ecological community.

FORAGE To wander in search of food or provisions; also the food or provisions found by this method.

FORAGING FLOCK Animals that join each other and move together while searching for food.

FORB An herb (a nonwoody plant, annual or perennial) that is not a grass.

FOREST CERTIFICATION Certification verifying that forests are well managed as defined by a particular standard and ensuring that certain wood or non-timber products come from responsibly managed forests.

FORESTRY The art and science of managing forests, tree plantations, and related natural resources for ecological or commercial values.

FOSSIL FUEL A hydrocarbon deposit, such as petroleum, coal, or natural gas, derived from living matter of a previous geologic time and used for fuel.

FRAGMENTATION The breaking up of something into smaller units. This often refers to habitat, which, when broken into smaller units, makes it harder for wildlife and plants to persist.

FRUIT The ripened ovary or ovaries of a seed-bearing plant, together with accessory parts, containing the seeds and variously modified to facilitate seed dispersal.

FRY A recently hatched fish that has fully absorbed its yolk sac and can now hunt and consume live food. The fry of live-bearers do not have yolk sacs and therefore need to begin feeding immediately after birth.

GASEOUS Existing in the state of a gas; not solid or liquid.

GEOGRAPHICAL RANGE The overall area within which a species can be found.

GEOMORPHOLOGIST A scientist who studies the shape of the Earth's land surface and how it is changed by rivers, mountains, oceans, air, and ice.

GEOTHERMAL ENERGY Energy from the internal heat of the Earth, extracted to produce power. Geothermal energy originates from the original formation of the planet, from radioactive decay of minerals, and from solar energy absorbed at the surface.

GILL The respiratory organ of aquatic animals that breathe oxygen dissolved in water. Fish, amphibians, insects, crustaceans, and mollusks may have gills.

GLACIAL ACTION The impacts that the presence and movement of a glacier have on the surrounding geology, including the shaping of the landscape through erosion and deposition of soil, minerals, rocks, and boulders and the compaction of the Earth's crust and mantle under the glacier's weight.

GLACIAL CYCLE Periods characterized by cooler and drier climates over most of the Earth and extension of large land and sea ice masses outward from the poles.

GLACIER A large sheet of ice formed in an alpine or other cool area.

GREENHOUSE EFFECT The heating of the surface of a planet or moon due to the presence of an atmosphere containing gases that absorb or reflect infra-red radiation.

GREENHOUSE GAS Any of the atmospheric gases that contribute to the greenhouse effect by absorbing infrared radiation produced by solar warming of the Earth's surface. They include carbon dioxide (CO_2), methane (CH_4), nitrous oxide (N_2O), and water vapor.

GROUNDWATER Water found beneath the surface of the Earth in saturated cracks and spaces in soil, sand, and rock.

GROUNDWATER-DEPENDENT ECOSYSTEM A network of ecological communities or species that are supported by or depend on groundwater for their survival.

GROUNDWATER–SURFACE WATER INTERACTION The flow of water from underground (groundwater) to the surface, from the surface water to underground, or both.

GRUB A soft, thick wormlike larva of an insect.

HABITAT The plants, animals, climate, topography, and other natural factors that compose the place where an organism or population lives.

HABITAT FRAGMENTATION A landscape-scale process involving the breaking up of large expanses of continuous habitat into smaller patches of discontinuous habitat.

HARMFUL ALGAL BLOOMS (HABS) Algae that grow massively, triggered by warm, nutrient-rich water, and deplete the water of oxygen and may produce toxins.

HEADWATERS The tributary farthest upstream from a river's outlet at a sea.

HEAVY METALS Metallic elements that have toxic effects in high concentrations, such as mercury, iron, copper, and lead.

HEMIMETABOLOUS Describing insects that undergo gradual development with no pupal stage. The young look like smaller versions of the adults, though often without wings. Showing incomplete metamorphosis.

HERBACEOUS Describing a plant with little or no woody tissue and that dies down to the soil level at the end of the growing season.

HERBACEOUS PERENNIALS Nonwoody plants that live for two or more growing seasons. The contrast is annuals, which live for only one growing season.

HERBIVORE An animal that eats mostly plants.

HETEROTROPH An organism that consumes other organisms in order to survive.

HIBERNATION A state of inactivity and metabolic depression in animals, characterized by lower body temperature, slower breathing, and lower metabolic rate.

HOLOMETABOLOUS Describing insects with complete metamorphosis, like bees, butterflies, and beetles. The young look like worms, go through a pupal stage, and then emerge as typical adults.

HORTICULTURE The industry and science of plant cultivation.

HYBRIDIZATION Crossbreeding of individuals from genetically different populations or species.

HYDRAULIC MINING A form of mining that employs extremely high-powered water cannons to dislodge rock material or move sediment.

HYDROELECTRIC ENERGY Electricity generated by the gravitational force of falling or flowing water.

HYDROGEN BOND A special type of chemical bond exhibited by molecules, notably water molecules, showing polarity and containing hydrogen.

HYDROLOGY The field of study relating to the Earth's water resources, including their movement, changes, and quality.

HYPHAE The strands of long cells that compose the living bodies of multicellular fungi.

IGNEOUS ROCK A rock type formed by the cooling and solidifying of molten magma. Igneous rocks can be either intrusive or extrusive.

IMPERVIOUS SURFACES Water-resistant materials, such as asphalt or concrete, that prevent the infiltration of rainwater or stormwater and can cause sheet flow.

INCOMPLETE METAMORPHOSIS Insect development, as in the grasshopper and cricket, in which the change is gradual and characterized by the absence of a pupal stage. See also *complete metamorphosis.*

INDIGENOUS Originating in and characteristic of a particular region or country; native.

INSECT An arthropod of the class Insecta that has three pairs of legs, a segmented body divided into three regions (head, thorax, and abdomen), one pair of antennae, and usually wings.

INTERGLACIAL A geological interval of warmer average global temperature that separates glacial periods within an ice age.

INTERNODES The sections of a plant stem between the nodes. The parts of the stem that do not produce leaves.

INTERTIDAL ZONE The region between the high-tide mark and the low-tide mark.

INTRUSIVE Describing igneous rock that cools and solidifies underground. Granite is an intrusive igneous rock. Contrast *extrusive.*

INVASIVE Describing any nonnative (exotic) species whose introduction causes or is likely to cause economic harm to the environmental or human health.

INVERTEBRATE An animal, such as an insect or mollusk, that lacks a backbone or spinal column.

JET STREAM Narrow, powerful bands of fast-moving wind in upper portions of the Earth's atmosphere on the boundaries between large air masses of differing temperatures.

KINETIC ENERGY The energy that an object possesses due to its motion.

LADDER FUELS Shrubs and trees that provide vertical continuity between strata, thereby allowing fire to move from surface fuels into the crowns of trees.

LAND SUBSIDENCE The sinking of the land surface level, often human-induced, such as through excessive groundwater pumping.

LANDFILL A burial site for the disposal of waste materials. It is usually lined to prevent leakage of fluids.

LARGE WOODY DEBRIS Fallen trees, logs, stumps, root wads, and branches in streams, lakes, and bays. Often defined as material that is greater than 4 inches in diameter.

LARVA The immature, wingless, feeding stage of an insect that undergoes complete metamorphosis.

LATITUDE The angular distance of a place north or south of the Earth's equator. On maps the lines parallel to the equator are lines depicting latitude.

LEACHATE Water that has percolated through a material or materials, such as those in a landfill, and extracted soluble compounds, sometimes rendering it toxic.

LEAF An aboveground plant organ specialized for photosynthesis.

LEAF LITTER Dead plant material, such as leaves, bark, needles, and twigs, that has fallen to the ground.

LEVEE A human-created wall or embankment intended to contain the flow of a river or other body of water and prevent flooding of the surrounding land.

LIANAS Woody vines that are rooted in the soil and use trees to climb to the canopy to access sunlight.

LICHEN An epiphytic photosynthetic organism formed by the symbiotic relationship between a fungus and an alga. Lichens grow in leaflike, crust-like, or branching forms on rocks and trees.

LIFE CYCLE The series of changes in the growth and development of an organism from its beginning as an independent life-form to its mature state in which offspring are produced.

LILIACEOUS Describing plants that are lilies or lily-like (Liliaceae, Agavaceae, Melanthiaceae, Nartheciaceae, Ruscaceae, Themidaceae, Tofieldiaceae).

LIMITING FACTOR A factor that controls a process, such as organism growth or species population, size, or distribution. The availability of food, predation pressure, and availability of shelter are examples of factors that could be limiting for an organism.

LITHOSPHERE The outer solid part of the Earth, including the crust and uppermost mantle.

LITTORAL ZONE The part of a lake, stream, or the ocean that is closest to the shore.

MACROINVERTEBRATE An invertebrate large enough to be seen with the unaided eye.

MAGMA Molten rock beneath the Earth's surface that is usually associated with tectonic faults or hot spot areas. When extruding to the surface, molten rock is known as lava.

MAMMAL Any vertebrate of the class Mammalia, having the body more or less covered with hair, nourishing the young with milk from the mammary glands, and giving birth to live young. Monotremes (the platypus and echidnas) are the exception, laying eggs.

MANAGED AQUIFER RECHARGE The intentional replenishment of aquifers for water storage or environmental benefit through strategies such as deep well injection, infiltration ponds, or flooding of agricultural fields.

MANTLE The portion of the Earth's interior between the core and the crust.

MARSUPIALS Mammals whose young are born undeveloped and grow outside their mother's body attached to one of her nipples. Marsupial females usually have pouches in which they carry their young.

MARSUPIUM The pouch or fold of skin on the abdomen of a female marsupial.

MECHANICAL ENERGY The sum of potential energy and kinetic energy present in the components of a mechanical system.

MEDITERRANEAN CLIMATE The two-season climate type found, roughly speaking, between 30 and 40 degrees latitude north and south of the equator, on the western side of continents; a climate having sunny, hot, dry summers and mild, rainy winters. Mediterranean climate areas are California, Chile, the Mediterranean, South Africa, and southwest Australia.

METAMORPHIC ROCK Rock formed when existing rock is subjected to high heat and pressure, transforming it into a new rock type.

METHANE An odorless, colorless, flammable gas that is the major constituent of natural gas and is used as a fuel and as an important source of hydrogen. Methane is a potent greenhouse gas.

MICROBE An organism that is microscopic (usually too small to be seen by the naked human eye).

MICROCLIMATE The climate in an area where atmospheric conditions differ from the surrounding area due to localized influences such as slope, aspect, and elevation.

MICROORGANISM Catchall term for any microscopic, unicellular life-form, including plankton, algae, bacteria, fungi, protists, and many others.

MIGRATORY Describing animals that regularly travel from one place to another, often over long distances.

MILLIPEDE Any terrestrial arthropod of the class Diplopoda, having a cylindrical body composed of 20 to more than 100 segments, each segment with two pairs of legs.

MOLECULE A group of two or more atoms linked together by sharing electrons in a chemical bond. Molecules are the fundamental components of chemical compounds and are the smallest parts of compounds that can participate in chemical reactions.

MOLT To periodically shed part or all of a coat or an outer covering, such as feathers, cuticle, or skin, which is then replaced by a new growth.

MONOTREMES Mammals that hatch young from eggs and have a single opening (a cloaca) for the digestive, urinary, and genital organs. Monotremes include only the platypus and the echidnas of Australia and New Guinea.

MORPHOLOGY The shape of a thing. Plant or animal morphology is the study of common forms and features that can identify a species. Geomorphology and river morphology involve the description of important land or stream features.

MOSS A spore-producing, nonvascular plant often growing in moist areas.

MUSHROOM The spore-producing reproductive structure of some fungi, often called a fruiting body and viewed as analogous to the fruit of a plant.

MYCELIUM The mass of hyphae that makes up the body of most fungi.

MYCORRHIZAL FUNGI The types of fungus that live in a mutually beneficial symbiotic relationship with plants, nurturing as well as being nurtured by their hosts.

NATIVE (PLANT OR ANIMAL) Being from a given environment without having been introduced by human activity.

NATURAL GAS A gas consisting primarily of methane that occurs naturally beneath the Earth's surface, often with or near petroleum deposits.

NATURALIST A person who observes or studies the world and all its forms and relationships, with an emphasis on observational rather than experimental methods.

NATURALIZED Describing any nonnative species reproducing without the aid of people.

NECTAR A sugar-rich liquid produced by plants, often to reward pollinators or guardians.

NICHE The lifestyle of a species; also the place a species lives, characterized by the biotic and abiotic factors that enable or constrain it.

NITROGEN OXIDES Binary compounds of oxygen and nitrogen. They are commonly found in vehicle exhaust, cigarette smoke, and smog. When dissolved in the atmosphere, they can lead to acid rain, which damages ecosystems.

NOCTURNAL Describing an animal that is active at night and usually not during the day.

NODE The place on a plant stem where buds, leaves, and branches originate.

NODULE A hollow swelling, often on a root, housing symbiotic bacteria that fix atmospheric nitrogen for the plant.

NONNATIVE Describing organisms that are not indigenous to the ecosystem to which they were introduced and that survive and reproduce without human intervention.

NONPOINT SOURCE (NPS) POLLUTION Pollution that comes from many disparate sources.

NONRENEWABLE RESOURCE A natural resource that cannot be produced, regrown, or regenerated on a scale that can sustain its consumption rate.

NONTIMBER FOREST PRODUCTS (NTFPS) Plants and animals people harvest in forests besides the boards from trees. NTFPs include deer, fish, honey, berries, greens, boughs, mushrooms, decorative items, and medicinal plants.

NOXIOUS WEED A plant species that has been designated by state or national agricultural authorities as a plant that is injurious to agricultural and/or horticultural crops and/or humans or livestock.

NUCLEUS The membrane-bound compartment within a cell in which DNA is stored.

NUTRIENT CYCLING The process by which mineral and gas nutrients are moved from the soil, water, or atmosphere to plants and then to animals and microorganisms and eventually back into the soil, water, or atmosphere.

NUTRIENT LOADING The contribution of large amounts of nutrients to an ecosystem, usually due to fertilizer or other chemical pollution runoff.

OLD-GROWTH FOREST A mature forest, characterized by large trees, well-developed structure, standing snags, and dead wood on the ground; a late-successional forest type for the area. It sometimes refers to undisturbed or never-harvested areas.

OLFACTORY Referring to the sense of smell. Olfactory nerves connect the smell organ (the nose) to the brain.

OLIGOTROPHIC Having a marked deficiency of nutrients or other materials needed to sustain life.

OMNIVORE An animal that eats both plant and animal material.

OPPOSITE LEAVES Leaves that arise from a single node in pairs on either side of a stem. See also *alternate leaves* and *whorled leaves*.

ORGANELLE A specialized part of a cell having some specific function; a cell organ. Examples are ribosomes, chloroplasts, and mitochondria.

OUTCROSSING Mating with someone other than yourself. Mixing genes by outcrossing is one of the primary methods of providing variability for natural selection to work on.

OUTLET The terminal end of a stream where it flows into the ocean or other body of water.

OVARY (PLANT) The ovule-bearing lower part of a pistil that ripens into a fruit.

OXBOW LAKE A crescent-shaped lake formed when a meander of a stream is cut off from the main channel.

OXYGEN An element that occurs as a diatomic gas (O_2) constituting 21 percent of the atmosphere by volume, combines with most elements, is essential for plant and animal respiration, and is required for nearly all combustion.

OZONE A colorless gas (O_3) that is formed naturally from atmospheric oxygen by electric discharge or exposure to ultraviolet radiation. It is also produced in the lower atmosphere by the photochemical reaction of certain pollutants. The ozone layer, 6 to 30 miles above the surface of the Earth, protects animals and plants from harmful UV radiation.

PALEO VALLEYS Highly permeable, ancient underground riverbeds.

PALEOENDEMIC Describing an endemic species with a fossil record of a more extensive geographical range than where it is currently found.

PARASITE An organism that lives in or on another, deriving benefit from its host at some cost to the host.

PARASITIC FUNGUS A type of fungus that lives in or on another organism, often weakening or killing its host.

PARENT MATERIAL The underlying material that soil develops from, generally bedrock that has decomposed in place, or material that has been deposited by wind, water, or ice.

PAROTID GLAND A salivary gland that in toads and some frogs and salamanders opens on the back, neck, and shoulder skin and produces distasteful secretions that may deter predators.

PARTICULATE MATTER Material suspended in the air in the form of minute solids or liquid droplets, especially when considered as an atmospheric pollutant.

PECTORAL FIN Either of the anterior pair of fins just behind the head of a fish, attached to the pectoral girdle, corresponding to the forelimbs of tetrapod vertebrates.

PENINSULAR RANGES Mountain ranges of the Sierra-Cascade system with a north-south axis and running from Riverside County in Southern California south through Baja California, Mexico.

PERENNIAL A plant that lives for more than 2 years. See also *annual*.

PERIPHYTON The mixture of algae, bacteria, other microorganisms, and organic detritus that coats underwater surfaces in a stream, lake, or other body of water.

PERMEABILITY The ease with which water moves through a particular soil or rock.

PETAL One of the often brightly colored parts of a flower surrounding the reproductive organs. Petals are attached to the receptacle underneath the carpels and stamens and may be separate or joined at their bases. Petals often function in visual attraction of pollinators.

PHOTOSYNTHESIS A process by which plants convert carbon dioxide into organic compounds, especially sugars, using the energy from sunlight.

PHYLOGENY The evolutionary development and history of a species or higher taxonomic grouping of organisms. Phylogeny looks at who descended from whom.

PISTIL The female, ovule-bearing organ of a flower, including the stigma, style, and ovary.

PLACENTAL Describing mammals having a placenta, that is, all mammals except monotremes.

PLANT COMMUNITY The group of plant species existing in a shared habitat or environment.

PLATE A section of the Earth's crust that moves about as a discrete whole, relative to other plates.

PLEISTOCENE The first epoch of the Quaternary period (about 2.6 million to 11,700 years ago). A time characterized by repeated glaciations and the emergence of humans.

POIKILOTHERMIC Describing an animal whose internal temperature varies along with that of the ambient environmental temperature. Most, but not all, ectotherms are poikilothermic.

POLLEN The male sex cells in plants. In flowering plants, pollen is produced in thin filaments in the flower, called stamens.

POLLINATION The first step in fertilization of a seed; the movement of pollen from male flower parts to female flower parts, often requiring wind or pollinators such as bees or other insects.

POLLINATOR The biotic agent (vector) that moves pollen from the anthers of a flower to the stigma to begin fertilization.

PRECIPITATION Condensed water vapor in the atmosphere that falls to Earth's surface, including rain, snow, and hail.

PREDATOR An organism that kills and eats other organisms (prey).

PRESCRIBED BURN ASSOCIATION A community-based network that collaborates to conduct prescribed burns for wildfire reduction, invasive species control, ecological health, and Tribal cultural purposes.

PRIMARY CONSUMER An animal that feeds on plants; an herbivore.

PROKARYOTES Single-celled organisms without a nucleus or other internal membrane-bound organelles. Examples include bacteria, cyanobacteria (blue-green algae), and organisms in the kingdom Archaebacteria.

PROTISTS A diverse group of eukaryotic microorganisms that are either unicellular or multicellular without specialized tissues.

PUPA The nonfeeding stage between the larva and adult in the metamorphosis of some insects, during which there is typically a complete transformation from larval to adult form within a protective cocoon or hardened case.

PUPATE Among insects, to go into a pupal stage, often wrapped in a cocoon or chrysalis, from which the insect will emerge as a fully formed adult.

RADIOACTIVE DECAY The process in which an unstable atomic nucleus spontaneously loses energy by emitting ionizing particles and radiation.

RAIN SHADOW The area of land that lies on the leeward (or downwind) side of a mountain range, thereby receiving less rainfall.

RAINFOREST A forested area where the annual rainfall is very high, often characterized by high levels of biodiversity.

REACH A segment of a stream, usually marked by geographic boundaries at the upstream and downstream ends.

RELICT A remnant left behind. A relictual species is the remnant of a formerly more widespread species or lineage. The Catalina ironwood is a relict of a once more-widespread species now restricted to the Channel Islands.

REPRODUCTION The sexual or asexual process by which organisms generate new individuals of the same kind.

RESERVOIR (WATER) A natural or artificial pond or lake used for the storage of water.

RESTORATION The process of assisting the recovery of a species, population, or ecosystem that has been degraded, damaged, or destroyed.

RIPARIAN Referring to the transition zone between a water body, such as a stream or lake, and an upland area. Riparian areas are often heavily vegetated with plants that are adapted to periodic flooding.

ROCK CYCLE The series of changes in which the three classes of rocks—igneous, sedimentary, and metamorphic—transform into one another by geological processes.

RUNOFF The flow of rain or snowmelt over land, typically when the land has become saturated or is impervious.

SALINITY The amount of salt present, especially dissolved salt in a body of water.

SALMONID The family of fish containing salmon, trout, whitefish, and others.

SAN ANDREAS FAULT A significant California earthquake fault running more than 800 miles from the Salton Sea to Cape Mendocino, formed by the relative horizontal movement of the Pacific and North American tectonic plates.

SAPROPHYTIC FUNGUS A type of fungus that eats dead trees, leaves, or other plant material, thereby contributing to the decomposition process.

SCAT Excrement, especially feces of an animal.

SCAVENGER A carnivorous animal that feeds on the tissues of dead animals.

SCLEROPHYLLOUS Describing leaves of trees and shrubs that are evergreen, hard, thick, leathery, and often small.

SEDIMENT Material that is carried in a suspended state by water or glacial action, ranging in size from microscopic particles of silt or clay up to rocks and boulders.

SEDIMENTARY ROCK Rock formed by the deposition of sediment out of its suspension in water or air, followed by a gradual solidification and hardening under pressure.

SEDIMENTATION The deposition of a stream's suspended load.

SELECTION PRESSURE The effect of selection on the relative frequency of one or more genes within a population.

SEPAL One of the usually green parts that surround and protect a flower bud and extend from the base of a flower after it has opened.

SERPENTINE SOIL A soil type characterized by a high magnesium-to-calcium ratio and generally low nutrient levels.

SHEET FLOW Water that flows over a large and irregular area, usually with a shallow depth, rather than in a restricted streambed.

SHRUB Category of woody plant, distinguished from a tree by its multiple stems and lower height, usually less than 15 to 20 feet tall.

SIERRA NEVADA The mountain range of eastern California running north-south from Plumas County to the area east of Bakersfield and comprising the eastern border of the Central Valley.

SILK Also known as gossamer, a protein fiber produced by spiders and many insects. It is often used to make webs or other structures that function as traps to capture prey, as nests or cocoons for protection for their offspring, or as capsules in which to undergo metamorphosis.

SILT A sedimentary material consisting of very fine particles intermediate in size between sand and clay.

SIMPLE LEAF A leaf that has an undivided blade. See also *compound leaf.*

SLOPE The steepness or angle of a hillside, usually in degrees.

SLOUGH A type of swamp or shallow lake system, typically formed as or by the backwater of a larger waterway.

SMOG Air pollution containing ozone and other reactive chemical compounds formed by the action of sunlight on nitrogen oxides and hydrocarbons, especially those in automobile exhaust.

SNAG A standing, partly or completely dead tree, often missing a top, bark, or branches.

SOIL A mixture of fine-particle mineral constituents, such as clay, silt, sand, and many trace minerals, along with decomposed organic matter, air, and water.

SOIL DEPOSITION The geological process by which material is added to a landform or landmass.

SOLAR POWER The useful power derived by converting the energy in sunlight into electricity.

SPAWN The mass of eggs deposited by fish, amphibians, mollusks, and crustaceans; also to lay these eggs.

SPECIALIST A species that requires a very specific habitat, food source, or other limiting factor in order to survive.

SPECIES A group of related organisms having common characteristics and capable of interbreeding.

SPINNERET Any of various tubular structures from which spiders and certain insect larvae, such as silkworms, secrete the silk threads from which they form webs or cocoons.

SPIRACLE An opening for "breathing" in the exoskeleton of insects, some spiders, and other arthropods.

STAMEN The pollen-producing organ of a flower, consisting of the filament and the anther.

STEM The supporting structure of a plant, which also serves to conduct and to store food materials.

STOMATE An opening (a pore) in the leaves or stems of plants, used for gas exchange.

SUBDUCTION ZONE A zone in which one crustal plate goes beneath another.

SUBSTRATE (ORGANISMS) The surface on or in which plants, algae, or certain animals, such as barnacles or clams, live or grow. A substrate may serve as a source of food for an organism or simply provide mechanical support.

SUBSTRATE (STREAMS) The material that rests at the bottom of a stream.

SUCCULENT Juicy or pulpy. Succulent plants often have thick, swollen stems or leaves to store water. This is not an evolutionary group; succulence has arisen in numerous plant lineages, such as cacti, lilies, ocotillo, and milkweeds.

SUSPENDED LOAD Sediment that is carried in a suspended state in a stream or other body of water. See also *bedload*.

SYMBIONT One member of a symbiotic relationship, often the smaller organism living inside another organism.

SYMBIOTIC Describing two organisms living closely together. Symbiotic relationships can be mutualistic, in which both organisms benefit; parasitic, in which one benefits and the other is harmed; or commensal, in which one benefits and the other is unaffected. Symbioses can also be obligatory (required) or facultative (optional).

SYRINX The vocal organ of a bird, consisting of thin vibrating muscles at or close to the division of the trachea into the bronchi.

TACTILE Of, pertaining to, endowed with, or affecting the sense of touch.

TAXA Units used in biological classification, such as kingdom, phylum, class, order, family, genus, or species; plural of *taxon*.

TECTONICS The processes that result in the structure and properties of the Earth's crust and its evolution through time; the motions of the Earth's plates and upper mantle.

TEMPERATE Describing the two geographical zones on the Earth's surface between the tropical zone and the Arctic or Antarctic Circle. Seasons are generally pronounced and variable in the temperate zones.

TENDRILS Ropelike extensions of plant leaves, branches, or flower stalks that coil and wrap around objects to allow the plant to climb and hold on.

TEPALS Plant parts that resemble petals but are not distinctly separated into sepals and petals; for instance, one may be in front and the other behind, but they look the same.

TERRESTRIAL Living predominantly or entirely on land.

TETRAPOD Vertebrate animal having four feet, legs, or leglike appendages or descended from four-limbed ancestors.

THERMAL ENERGY A form of energy that manifests itself as an increase in temperature.

THORAX The middle division of the body of an insect, to which the wings and legs are attached. The thorax lies between the head and the abdomen.

THREATENED Describing any species or community (including animals, plants, or fungi) that are vulnerable to extinction in the near future.

TIDAL POWER A form of hydropower that converts the energy of tides into electricity or other useful forms of power.

TOPOGRAPHY The shape of the land. The terrain in a given area, including hills, mountains, valleys, cliffs, and rivers.

TORPOR A (usually short-term) state of decreased physiological activity in an animal, usually characterized by a reduced body temperature and rate of metabolism.

TRANSFORM PLATE BOUNDARY A tectonic boundary where two plates slide horizontally past one another in opposite directions.

TRANSMONTANE CALIFORNIA Land to the east of the Sierra-Cascade mountains.

TRANSVERSE RANGES The mostly east-west mountain ranges north of the Los Angeles basin.

TRIBUTARY A stream that flows into another, usually larger, stream.

TROPHIC LEVEL The levels of a food chain, such as producers, primary consumers, secondary consumers, and tertiary consumers.

TROPIC, TROPICAL Pertaining to the tropics, the equatorial area between the Tropic of Cancer and the Tropic of Capricorn.

UNDERSTORY The layer of vegetation beneath the canopy.

UPWELLING A movement of deeper water to the surface. Upwelling occurs when winds push surface water away, allowing deeper nutrient-rich water to rise to the ocean's surface.

VERNAL POOL A seasonal body of standing water that fills in the wet season and dries completely from June to November. Vernal pools are free of fish, can support unique plant assemblages, and provide important breeding habitat for many terrestrial or semiaquatic species such as frogs, salamanders, and turtles.

VERTEBRATE An animal that has a spinal column.

VOLATILE ORGANIC COMPOUNDS Compounds that have a high vapor pressure and low water solubility. Many are human-made chemicals from the manufacture of paints, adhesives, petroleum products, fuels, solvents, pharmaceuticals, and refrigerants. They are a common source of contamination because many are toxic and are known or suspected human carcinogens.

WATER CYCLE The continuous process in which the Earth's water changes form between gas, liquid, and solid and is moved around the surface and atmosphere of the planet. It is also known as the hydrologic cycle.

WATERSHED The topographic region drained by a river and its tributaries. Also called a drainage basin.

WAVE POWER The transport of energy by ocean surface waves, and the capture of that energy to do useful work—for example, for electricity generation, water desalination, or the pumping of water into reservoirs.

WETLANDS Land areas that are permanently or seasonally inundated with water.

WHORLED LEAVES An arrangement of leaves that forms a circular pattern, with multiple leaves emerging from a node. See also *alternate leaves* and *opposite leaves*.

WOODY PLANT A plant having hard lignified tissues or woody parts, especially stems.

The definitions in the glossary were derived from a variety of sources, including government documents (from the US Geological Survey, US Department of Agriculture, Natural Resource Conservation Service, and others), academic course sites, and online dictionaries, such as Wikipedia.

Index

Founded in 1893,
UNIVERSITY OF CALIFORNIA PRESS
publishes bold, progressive books and journals
on topics in the arts, humanities, social sciences,
and natural sciences—with a focus on social
justice issues—that inspire thought and action
among readers worldwide.

The UC PRESS FOUNDATION
raises funds to uphold the press's vital role
as an independent, nonprofit publisher, and
receives philanthropic support from a wide
range of individuals and institutions—and from
committed readers like you. To learn more, visit
ucpress.edu/supportus.